MECHANICS AND RELATIVITY

MECHANICS AND RELATIVITY

BY

ETHAN BOOTH
EVIE EDWARDS
DENAYE HUGHES

KNOWLEDGE
BAKERS

KNOWLEDGE BAKERS

ISBN: 978-81-9755-941-9 (HB)

Published by : **Knowledge Bakers**
Pune, India
Email: info@knowledgebakers.com

PREFACE

Mechanics and relativity stand as foundational pillars in the realm of physics, offering profound insights into the behavior of objects in motion and the fundamental nature of space, time, and gravity. Mechanics, with its classical principles established by Newton, provides a framework for understanding the motion of macroscopic objects, from the orbits of planets to the trajectories of projectiles. Relativity, introduced by Einstein in the early 20th century, revolutionized our understanding of the universe, revealing the intricate interplay between space, time, and energy on both cosmic and microscopic scales. As we embark on this exploration of mechanics and relativity, we delve into a realm where the laws of physics govern the behavior of the universe, from the smallest particles to the largest galaxies, uncovering the mysteries of motion, gravity, and the fabric of spacetime.

This book, titled "Mechanics and Relativity," is crafted to provide a comprehensive overview of these two fundamental branches of physics. Through collaboration with leading physicists, researchers, and educators in the fields of mechanics and relativity, we aim to offer readers a multidimensional and insightful journey through the intricacies of classical and modern physics.

As editors of this volume, we have curated a collection of chapters that cover a wide range of topics in mechanics and relativity, including classical mechanics, celestial mechanics, special relativity, general relativity, and cosmology. Each chapter is designed to provide readers with a deep understanding of the fundamental principles, mathematical formalism, experimental evidence, and theoretical implications of mechanics and relativity.

Furthermore, this book seeks to highlight the profound implications of mechanics and relativity for our understanding of the universe and the nature of reality. By exploring concepts such as spacetime curvature, time dilation, gravitational waves, and the expanding universe, we aim to inspire readers to contemplate the deep connections between physics, philosophy, and the human quest for knowledge and understanding.

As we embark on this enlightening journey through the world of mechanics and relativity, we extend our gratitude to the contributors whose expertise and dedication have enriched this book. We hope that "Mechanics and Relativity" serves as a valuable resource for students, educators, researchers, and enthusiasts alike, empowering them to deepen their understanding of the fundamental laws that govern the cosmos and to appreciate the beauty and elegance of the universe as revealed by the principles of mechanics and relativity.

Ethan Booth
Evie Edwards
Denaye Hughes

Contents

I

CLASSICAL MECHANICS

1

Introduction to Classical Mechanics

Classical mechanics is the study of the motion of bodies under the action of physical forces. A *force* is any influence that can cause an object to change its velocity. The object can be anything from an elementary particle to a galaxy. Of course anything larger than an elementary particle is ultimately a composite of elementary particles, but fortunately we usually don't have to consider all those, and can *coarse-grain* to the scale of the objects at hand. As is true for any physical model, classical mechanics is an approximation and has its limits - it breaks down at very small scales, high speeds and large gravitational fields - but within its range of applicability (which includes pretty much every single phenomenon in everyday life) it is extremely useful.

Classical mechanics is based on a small number of *physical laws*, which are mathematical formulations of a physical observation. Some laws can be derived from others, but you cannot derive all of them from scratch. Some laws are *axioms*, and we'll assume they are valid. The laws we'll encounter can be divided up in three classes: Newton's laws of motion, conservation laws and force laws. As we'll see, the three conservation laws of classical mechanics (of energy, momentum and angular momentum) can be derived from Newton's second and third laws of motion, as can Newton's first law. The force laws give us the force exerted by a certain physical system - a compressed spring (Hooke's law) or two charged particles (Coulomb's law) for example. These also feed back into Newton's laws of motion, although they cannot be derived from these and are axioms by themselves.

In addition to the physical laws, there is a large number of *definitions* - which should not be confused with the laws. Definitions are merely convenient choices. A good example is the definition of the number π: half the ratio of the circumference to the radius of a circle. As you have no doubt noticed, it is very convenient that this number has gotten its own symbol that is universally recognized, as it pops up pretty much everywhere. However, there is no axiom here, as we are simply taking a ratio and giving it a name.

1.1 Dimensions and units

In physics in general, we are interested in relating different *physical quantities* to one another - we want to answer questions like 'how much work do I need to do to get this box up to the third floor'? In order to be able to give an answer, we need certain measurable quantities as input - in the present case, the mass of the box and the height of a floor. Then, using our laws of physics, we will be able to produce another measurable quantity as our answer - here the amount of work needed. Of course, you could check this answer, and thus validate our physical model of reality, by measuring the quantity in question.

Measurable (or 'physical', or 'observational') quantities aren't just numbers - the fact that they correspond to something physical matters, and 10 seconds is something very different from 10 meters, or 10 kilograms. The term we use to express this is, rather unfortunately, to say that physical quantities have a *dimension* - not to be confused with length, height and width. Anything that has a dimension can be measured, and to do so we use *units* - though there may be different units in which we measure the same quantity, such as centimeters and inches for length. When measuring the same quantity in different units, you can always convert

[1] At the time of writing, the unit of mass is still determined using a prototype in Paris, however, a redefined unit based on the value of Planck's constant is expected to be adopted on May 20, 2019.

quantity	symbol	unit	symbol	based on
length	L	meter	m	speed of light
time	T	second	s	caesium atom oscillation
mass	M	kilogram	kg	Planck's constant[1]
current	I	Ampère	A	electron charge
temperature	T	Kelvin	K	Boltzmann's constant
luminosity	J	candela	cd	monochromatic radiation
particle count	N	mole	mol	Avogadro's constant

Table 1.1: Overview of the SI quantities and units, and the physical constants they are (or are proposed to be) based on.

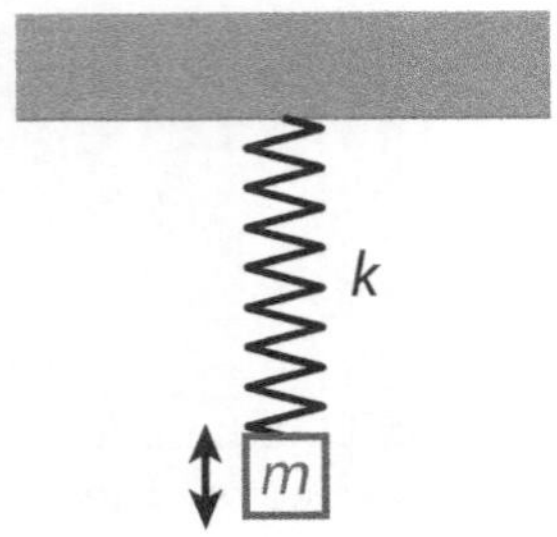

Figure 1.1: A harmonic oscillator: a mass m suspended on a spring with spring constant k, oscillating with a frequency ω.

between them - there are 2.54 centimeters in an inch - but it's meaningless to try to convert centimeters into seconds, because length and time are different quantities - they have different dimensions.

We will encounter only three different basic quantities, which have the dimensions of length (L), time (T), and mass (M). Thanks to the Napoleonic conquest of Europe in the early 1800s, we have a basic unit for each of these: meters (m) for length, seconds (s) for time, and kilograms (kg) for mass. Although we won't encounter them here, the standard system of units (called the Système International, or SI) has four more of these basic pairs: (electric) current I, measured in Ampères (A), temperature T, measured in Kelvin (K), luminosity J, measured in candelas (cd), and 'amount of stuff', measured in moles (mol), see table 1.1. Unfortunately, although this system is commonly used in (continental) Europe and in many other parts of the world, it is not everywhere, notably in the US, where people persist in using such things as inches and pounds, so you'll often have to convert between units.

From the seven basic quantities in the SI, all others can be derived. For example, speed is defined as the distance traveled (length) divided by the time it took, so speed has the dimension of L/T and is measured in units of m/s. Note that in order to be able to compare two quantities, they must have the same dimension. This simple observation has an important consequence: in any physics equation, the dimensions on both sides of the equality sign always have to be the same. There's no bargaining on this point: equating two quantities with different dimensions does not make any kind of sense, so if you find that that's what you're doing at any point, backtrack and find where things went wrong.

1.2 DIMENSIONAL ANALYSIS

Although you will of course need a complete physical model (represented as a set of mathematical equations) to fully describe a physical system, you can get surprisingly far with a simple method that requires no detailed knowledge at all. This method is known as *dimensional analysis*, and based on the observation in the previous section that the two sides of any physical equation have to have the same dimension. You can use this principle to qualitatively understand a system, and make predictions on how it will respond quantitatively if you change some parameter. To understand how dimensional analysis works, an example is probably the most effective - we'll take one that is ubiquitous in classical mechanics: a mass oscillating on a spring (known as the *harmonic oscillator*), see figure 1.1.

1.2.1 WORKED EXAMPLE: DIMENSIONAL ANALYSIS OF THE HARMONIC OSCILLATOR

Consider the harmonic oscillator consisting of a mass of magnitude m, suspended on a spring with spring constant k. If you pull down the mass a bit and release, it will oscillate with a frequency ω. Can we predict how this frequency will change if we double the mass?

There are two ways to answer this question. One is to consider all the forces acting on the mass, then use Newton's second law to derive a differential equation (known as the equation of motion) for the mass, solve it, and from the solution determine what happens if we change the mass. The second is to consider the dimensions of the quantities involved. We have a mass, which has dimension of mass (M), as it is one of our basic quantities. We have a spring with spring constant k, which has dimensions of force per unit length, or mass per unit time squared:

$$[k] = F/L = MLT^{-2}/L = M/T^2. \tag{1.1}$$

Note the notation $[k]$ for the dimension of k. For the frequency, we have $[\omega] = 1/T$. Now we know that the frequency is a function of the spring constant and the mass, and that both sides of that equation must have the same sign. Since there is no mass in the dimension of the frequency, but it exists in the dimension of both the spring constant and the mass, we know that ω must depend on the ratio of k and m: $\omega \sim k/m$. Now $[k/m] = 1/T^2$, and from $[\omega] = 1/T$, we conclude that we must have

$$\omega \sim \sqrt{k/m}. \tag{1.2}$$

Equation (1.2) allows us to answer our question immediately: if we double the mass, the frequency will decrease by a factor $\sqrt{2}$.

Note that in equation (1.2) I did not write an equals sign, but a 'scales as' sign ($\sim$, sometimes also written as $\propto$). That is because dimensional analysis will not tell us about any numerical factor that may appear in the expression, as those numerical factors have no unit (or, more correctly, have no dimension - they are *dimensionless*).

You may object that there might be another factor at play: shouldn't gravity matter? The answer is no, as we can also quickly see from dimensional analysis. The force of gravity is given by mg, introducing another parameter g (the gravitational acceleration) with dimension $[g] = L/T^2$. Now if the frequency were to depend on g, there has to be another factor to cancel the dependence on the length, as the frequency itself is length-independent. Neither m nor k has a length-dependence in its dimension, and so they *cannot* 'kill' the L in the dimension of g; the frequency therefore also cannot depend on g - which we have now figured out without invoking any (differential) equations!

Above, I've sketched how you can use dimensional analysis to arrive at a physical scaling relation through inspection: we've combined the various factors to arrive at the right dimension. Such combinations are not always that easy to see, and in any case, you may wonder if you've correctly spotted them all. Fortunately, there is a more robust method, that we can also use to once again show that the frequency is independent of the gravitational acceleration. Suppose that in general ω could depend on k, m and g. The functional dependence can then be written as[2]

$$[\omega] = [k^\alpha m^\beta g^\gamma] = (M/T^2)^\alpha M^\beta (L/T^2)^\gamma = M^{\alpha+\beta} T^{-2(\alpha+\gamma)} L^\gamma, \tag{1.3}$$

which leads to three equations for the exponents:

$$\begin{aligned} \alpha + \beta &= 0, \\ -2(\alpha + \gamma) &= -1, \\ \gamma &= 0, \end{aligned}$$

which you can easily solve to find $\alpha = 1/2$, $\beta = -1/2$, $\gamma = 0$, which gives us equation (1.2). This method[3] will allow you to get dimensional relations in surprisingly many different cases, and is used by most physicist as a first line of attack when they first encounter an unknown system.

[2]The actual function may of course contain multiple terms which are summed, but all those must have the same dimension. Operators like sines and exponentials must be dimensionless, as there are no dimensions of the form $\sin(M)$ or e^L. The only allowable dimensional dependencies are thus power laws.

[3]The method is sometimes referred to as the Rayleigh algorithm, after John William Strutt, Lord Rayleigh (1842-1919), who applied it, among other things, to light scattering in the air. The result of Rayleigh's analysis can be used to explain why the sky is blue.

1.3 PROBLEMS

1.1 **Harmonic oscillator revisited** Suppose you have a small object of mass m, which you attach to a spring of spring constant k (which itself is fixed to a wall at its other end, figure 1.1). Above, we derived an expression for the frequency of oscillation of the mass. We also argued that it should be the same for both a horizontally-positioned and a vertically-positioned oscillator, i.e., that the frequency is independent of the gravitational acceleration g.

(a) Show that the frequency of oscillation is also independent of its amplitude A (the maximum distance from the equilibrium position the oscillating mass reaches).

(b) Use dimensional analysis to derive an expression for the maximum velocity of the mass during the oscillation, as a function of m, k, and A.

1.2 In physics, we assume that quantities like the speed of light (c) and Newton's gravitational constant (G) have the same value throughout the universe, and are therefore known as physical constants. A third such constant from quantum mechanics is Planck's constant ($\hbar$, an h with a bar). In high-energy physics, people deal with processes that occur at very small length scales, so our regular SI-units like meters and seconds are not very useful. Instead, we can combine the fundamental physical constants into different basis values.

(a) Combine c, G and $\hbar$ into a quantity that has the dimensions of length.

(b) Calculate the numerical value of this length in SI units (this is known as the Planck length). You can find the numerical values of the physical constants in appendix B.

(c) Similarly, combine c, G and $\hbar$ into a quantity that has the dimensions of energy (indeed, known as the Planck energy) and calculate its numerical value.

1.3 **Reynolds numbers** Physicists often use *dimensionless quantities* to compare the magnitude of two physical quantities. Such numbers have two major advantages over quantities with numbers. First, as dimensionless quantities carry no units, it does not matter which unit system you use, you'll always get the same value. Second, by comparing quantities, the concepts 'big' and 'small' are well-defined, unlike for quantities with a dimension (for example, a distance may be small on human scales, but very big for a bacterium). Perhaps the best known example of a dimensionless quantity is the *Reynolds number* in fluid mechanics, which compares the relative magnitude of inertial and drag forces acting on a moving object:

$$\mathrm{Re} = \frac{\text{inertial forces}}{\text{drag forces}} = \frac{\rho v L}{\eta}, \tag{1.4}$$

where ρ is the density of the fluid (either a liquid or a gas), v the speed of the object, L its size, and η the viscosity of the fluid. Typical values of the viscosity are 1.0 mPa·s for water, 50 mPa·s for ketchup, and 1.0 μPa·s for air.

(a) Estimate the typical Reynolds number for a duck when flying and when swimming (you may assume that the swimming happens entirely submerged). NB: This will require you looking up or making educated guesses about some properties of these birds in motion. In either case, is the inertial or the drag force dominant?

(b) Estimate the typical Reynolds number for a swimming bacterium. Again indicate which force is dominant.

(c) Oil tankers that want to make port in Rotterdam already put their engines in reverse halfway across the North sea. Explain why they have to do so.

(d) Express the Reynolds number for the flow of water through a (circular) pipe as a function of the radius R of the pipe, the volumetric flow rate (i.e., volume per second that flows through the pipe) Q, and the kinematic viscosity $\nu \equiv \eta/\rho$.

(e) For low Reynolds number, fluids will typically exhibit so-called laminar flow, in which the fluid particles all follow paths that nicely align (this is the transparent flow of water from a tap at low flux). For higher Reynolds number, the flow becomes turbulent, with many eddies and vortices (the white-looking flow of water from the tap you observe when increasing the flow rate). The maximum Reynolds number for which the flow is typically laminar is experimentally measured to be about 2300. Estimate the flow velocity and volumetric flow rate of water from a tap with a 1.0 cm diameter in the case that the flow is just laminar.

1.4 The *escape velocity* of a planet is defined as the minimal initial velocity an object must have to escape its gravitational pull completely (and thus go fast enough to defy the rule that 'what goes up must come down').

(a) From Newton's universal law of gravitation (equation 2.9), determine the dimension of the gravitational constant G.

(b) Use dimensional analysis to show that for a planet of mass M and radius R, the escape velocity scales as $v \sim \sqrt{MG/R}$.

(c) A more detailed calculation shows that in fact we have $v_{\text{escape}} = \sqrt{2GM/R}$. Express this value of the escape velocity in terms of the (mass) density ρ of the planet, instead of its mass M.

(d) The average density of the moon is about 6/10th that of the Earth, and the Moon's radius is about 11/40 times that of the Earth. From these numbers and your answer at (c), calculate the ratio of the escape velocities of the Moon and the Earth, and explain why the Apollo astronauts needed a huge rocket to get to the Moon, and only a tiny one to get back.

2

FORCES

2.1 NEWTON'S LAWS OF MOTION

As described in chapter 1, classical mechanics is based on a set of axioms, which in turn are based on (repeated) physical observations. In order to formulate the first three axioms, we will need to first define three quantities: the (instantaneous) velocity, acceleration and momentum of a particle. If we denote the position of a particle as $\boldsymbol{x}(t)$ - indicating a vector[1] quantity with the dimension of length that depends on time, we define its velocity as the time derivative of the position:

$$\boldsymbol{v}(t) = \dot{\boldsymbol{x}}(t) = \frac{\mathrm{d}\boldsymbol{x}(t)}{\mathrm{d}t}. \tag{2.1}$$

Note that we use an overdot to indicate a *time* derivative, we will use this convention throughout these notes. The acceleration is the time derivative of the velocity, and thus the second derivative of the position:

$$\boldsymbol{a}(t) = \ddot{\boldsymbol{x}}(t) = \frac{\mathrm{d}\boldsymbol{v}(t)}{\mathrm{d}t} = \frac{\mathrm{d}^2\boldsymbol{x}(t)}{\mathrm{d}t^2}. \tag{2.2}$$

Finally the *momentum* of a particle is its mass times its velocity:

$$\boldsymbol{p}(t) = m\boldsymbol{v}(t) = m\dot{\boldsymbol{x}}(t). \tag{2.3}$$

We are now ready to give our next three axioms. You may have encountered them before; they are known as Newton's three laws of motion.

Axiom 1 (Newton's first law of motion)**.** *As long as there is no external action, a particle's velocity will remain constant.*

Note that the first law includes particles at rest, i.e., $\boldsymbol{v} = 0$. We will define the general 'external action' as a *force*, therefore a force is now anything that can change the velocity of a particle. The second law quantifies the force.

Axiom 2 (Newton's second law of motion)**.** *If there is a net force acting on a particle, then its instantaneous change in momentum due to that force is equal to that force:*

$$\boldsymbol{F}(t) = \frac{\mathrm{d}\boldsymbol{p}(t)}{\mathrm{d}t}. \tag{2.4}$$

Now since $\boldsymbol{p} = m\boldsymbol{v}$ and $\boldsymbol{a} = \mathrm{d}\boldsymbol{v}/\mathrm{d}t$, if the mass is constant we can also write (2.4) as $\boldsymbol{F} = m\boldsymbol{a}$, or

$$\boldsymbol{F}(t) = m\ddot{\boldsymbol{x}}(t), \tag{2.5}$$

which is the form in we will use most. Based on the second law, we see that a force has the physical dimension of a mass times a length divided by a time squared - since this is quite a lot to put in every time, we

[1]Appendix A.1 lists some basic properties of vectors that you may find useful.

Isaac Newton (1642-1727) was a British physicist, astronomer and mathematician, who is widely regarded as one of the most important scientists in history. Newton was a professor at Cambridge from 1667 till 1702, where he held the famous Lucasian chair in mathematics. Newton invented infinitesimal calculus to be able to express the laws of mechanics that now bear his name (section 2.1) in mathematical form. He also gave a mathematical description of gravity (equation 2.9), from which he could derive Kepler's laws of planetary motion (section 6.4). In addition to his work on mechanics, Newton made key contributions to optics and invented the reflection telescope, which uses a mirror rather than a lens to gather light. Having retired from his position in Cambridge, Newton spend most of the second half of his life in London, as warden and later master of the Royal mint, and president of the Royal society.

Figure 2.1: Portrait of Isaac Newton by Godfrey Kneller (1689) [2].

define the dimension of force as such: $F = MLT^{-2}$. Likewise, we define a unit, the Newton (N), as a kilogram times a meter per second squared: $\mathrm{N} = \mathrm{kg}\cdot\mathrm{m/s}^2$. Therefore, in principle Newton's second law of motion can also be used to measure forces, though we will often use it the other way around, and calculate changes in momentum due to a known force.

Note how Newton's first law follows from the second: if the force is zero, there is no change in momentum, and thus (assuming constant mass) a constant velocity. Note also that although the second law gives us a quantification of the force, by itself it will not help us achieve much, as we at present have no idea what the force is (though you probably have some intuitive ideas from experience) - for that we will use the force laws of the next section. Before we go there, there is another important observation on the nature of forces in general.

Axiom 3 (Newton's third law of motion). *If a body exerts a force $\boldsymbol{F}_1$ on a second body, the second body exerts an equal but opposite force $\boldsymbol{F}_2$, on the first, i.e., the forces are equal in magnitude but opposite in direction:*

$$\boldsymbol{F}_1 = -\boldsymbol{F}_2. \tag{2.6}$$

2.2 FORCE LAWS

Newton's second law of motion tells us what a force *does*: it causes a change in momentum of any particle it acts upon. It does not tell us where the force comes from, nor does it care - which is a very useful feature, as it means that the law applies to all forces. However, we do of course need to know what to put down for the force, so we need some rule to determine it independently. This is where the force laws come in.

2.2.1 SPRINGS: HOOKE'S LAW

One very familiar example of a force is the spring force: you need to exert a force on something to compress it, and (in accordance with Newton's third law), if you press on something you'll feel it push back on you. The simplest possible object that you can compress is an ideal spring, for which the force necessary to compress it scales linearly with the compression itself. This relation is known as **Hooke's law**:

$$\boldsymbol{F}_s = -k\boldsymbol{x}, \tag{2.7}$$

where $\boldsymbol{x}$ is now the displacement (from rest) and k is the spring constant, measured in newtons per meter. The value of k depends on the spring in question - stiffer springs having higher spring constants.

Hooke's law gives us another way to *measure* forces. We have already defined the unit of force using Newton's second law of motion, and we can use that to calibrate a spring, i.e., determine its spring constant, by determining the displacement due to a known force. Once we have k, we can simply measure forces by measuring displacements - this is exactly what a spring scale does.

Robert Hooke (1635-1703) was a British all-round natural scientist and architect. He discovered the force law named after him in 1660, which he published first as an anagram: 'ceiiinosssttuv', so he could claim the discovery without actually revealing it (a fairly common practice at the time); he only provided the solution in 1678: 'ut tensio, sic vis' ('as the extension, so the force'). Hooke made many contributions to the development of microscopes, using them to reveal the structure of plants, coining the word cell for their basic units. Hooke was the curator of experiments of England's Royal Society for over 40 years, combining this position with a professorship in geometry and the job of surveyor of the city of London after the great fire of 1666. In the latter position he got a strong reputation for a hard work and great honesty. At the same time, he was frequently at odds with his contemporaries Isaac Newton and Christiaan Huygens; it is not unlikely that they independently developed similar notions on, among others on the inverse-square law of gravity.

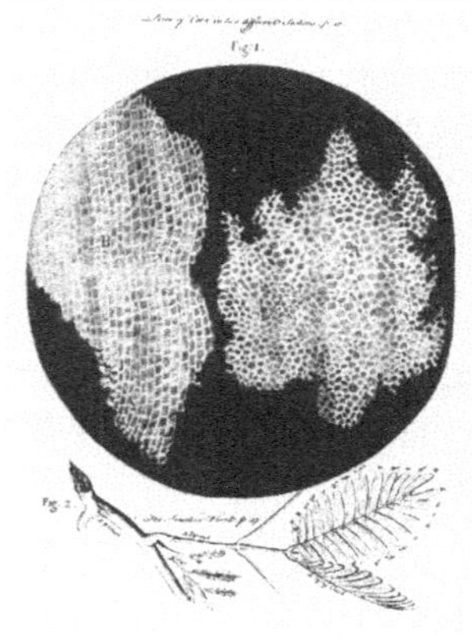

Figure 2.2: Drawing of the cell structure of cork by Hooke, from his 1665 book Micrographia [3]. No portraits of Hooke survive.

2.2.2 GRAVITY: NEWTON'S LAW OF GRAVITY

A second and probably even more familiar example is force due to gravity, at the local scale, i.e., around you, in the approximation that the Earth is flat. Anything that has mass attracts everything else that has mass, and since the Earth is very massive, it attracts all objects in the space around you, including yourself. Since the force of gravity is weak, you won't feel the pull of your book, but since the Earth is so massive, you do feel its pull. Therefore if you let go of something, it will be accelerated towards the Earth due to its attracting gravitational force. As demonstrated by Galilei (and some guys in spacesuits on a rock we call the moon[2]), the acceleration of any object due to the force of gravity is the same, and thus the force exerted by the Earth on any object equals the mass of that object times this acceleration, which we call $\boldsymbol{g}$:

$$\boldsymbol{F}_g = m\boldsymbol{g}. \tag{2.8}$$

Because the Earth's mass is not exactly uniformly distributed, the magnitude of $\boldsymbol{g}$ varies slightly from place to place, but to good approximation equals $9.81\,\mathrm{m/s^2}$. It always points down.

Although equation (2.8) for local gravity is handy, its range of application is limited to everyday objects at everyday altitudes - say up to a couple thousand kilograms and a couple kilometers above the surface of the Earth, which is tiny compared to Earth's mass and radius. For larger distances and bodies with larger mass - say the Earth-Moon, or Earth-Sun systems - we need something else, namely **Newton's law of gravitation** between two bodies with masses m_1 and m_2 and a distance r apart:

$$\boldsymbol{F}_\mathrm{G} = -G\frac{m_1 m_2}{r^2}\hat{\boldsymbol{r}}, \tag{2.9}$$

where $\hat{\boldsymbol{r}}$ is the unit vector pointing along the line connecting the two masses, and the proportionality constant $G = 6.67 \cdot 10^{-11}\ \mathrm{N \cdot m^2/kg^2}$ is known as the gravitational constant (or Newton's constant). The minus sign indicates that the force is attractive. Equation (2.9) allows you to actually calculate the gravitational pull that your book exerts on you, and understand why you don't feel it. It also lets you calculate the value of g - simply fill in the mass and radius of the Earth. If you wish to know the value of g on any other celestial body, you can put in its particulars, and compare with Earth. You'll find you'd 'weigh' 3 times less on Mars and 6 times less on the Moon. Most of the time we can safely assume the Earth is flat and use (2.8), but in particular for celestial mechanics and when considering satellites we'll need to use (2.9).

[2]To be precise, astronaut David Scott of the Apollo 15 mission in 1971, who dropped both a hammer and a feather and saw them fall at exactly the same rate, as shown in this NASA movie.

Galileo Galilei (1564-1642) was an Italian physicist and astronomer, who is widely regarded as one of the founding figures of modern science. Unlike classical philosophers, Galilei championed the use of experiments and observations to validate (or disprove) scientific theories, a practice that is the cornerstone of the scientific method. He pioneered the use of the telescope (newly invented at the time) for astronomical observations, leading to his discovery of mountains on the moon an the four largest moons of Jupiter (now known as the Galilean moons in his honor). On the theoretical side, Galilei argued that Aristotle's argument that heavy objects fall faster than light ones is incorrect, and that the acceleration due to gravity is equal for all objects (equation 2.8). Galilei also strongly advocated the heliocentric worldview introduced by Copernicus in 1543, as opposed to the widely-held geocentric view. Unfortunately, the Inquisition thought otherwise, leading to his conviction for heresy with a sentence of life-long house arrest in 1633, a position that was only recanted by the church in 1995.

Figure 2.3: Portrait of Galileo Galilei by Justus Sustermans (1636) [4].

2.2.3 Electrostatics: Coulomb's law

Like two masses interact due to the gravitational force, two charged objects interact via Coulomb's force. Because charge has two possible signs, Coulomb's force can both be attractive (between opposite charges) and repulsive (between identical charges). Its mathematical form strongly resembles that of Newton's law of gravity:

$$\boldsymbol{F}_\mathrm{C} = k_e \frac{q_1 q_2}{r^2} \hat{\boldsymbol{r}}, \tag{2.10}$$

where q_1 and q_2 are the signed magnitudes of the charges, r is again the distance between them, and $k_\mathrm{e} = 8.99 \cdot 10^9\ \mathrm{N \cdot m^2/C^2}$ is Coulomb's constant. For everyday length and force scales, Coulomb's force is much larger than the force of gravity.

2.2.4 Friction and drag

Why did it take the genius of Galilei and Newton to uncover Newton's first law of motion? Because everyday experience seems to contradict it: if you don't exert a force, you won't keep moving, but gradually slow down. You know of course why this is: there's *drag* and *friction* acting on a moving body, which is why it's much easier (though not necessarily handier) for a car to keep moving on ice than on a regular tarmac road (less friction on ice), and why walking through water is so much harder than walking through air (more drag in water). The *medium* in which you move can exert a drag force on you, and the *surface* over which you move exerts friction forces. These of course are the forces responsible for slowing you down when you stop exerting force yourself, so the first law doesn't apply, as there are forces acting.

For low speeds, the drag force typically scales linearly with the velocity of the moving object. Drag forces for objects moving through a (fluid) medium moreover depend on the properties of the medium (its viscosity η) and the cross-sectional area of the moving object. For a sphere of radius R moving at velocity $\boldsymbol{v}$, the drag force is given by **Stokes' law**:

$$\boldsymbol{F}_\mathrm{d} = -6\pi\eta R \boldsymbol{v}. \tag{2.11}$$

The more general version for an object of arbitrary shape is $\boldsymbol{F}_d = \zeta \boldsymbol{v}$, where ζ is a proportionality constant. Stokes' law breaks down at high velocities, for which the drag force scales quadratically with the speed:

$$F_\mathrm{d} = \frac{1}{2} \rho c_\mathrm{d} A v^2, \tag{2.12}$$

where ρ is the density of the fluid, A the cross-sectional area of the object, v it's speed, and c_d it's dimensionless drag coefficient, which depends on the object's shape and surface properties. Typical values for the drag coefficient are 1.0 for a cyclist, 1.2 for a running person, 0.48 for a Volkswagen Beetle, and 0.19 for a modern aerodynamic car. The direction of the drag force is still opposite that of the motion.

Frictional forces are due to two surfaces sliding past each other. It should come as no surprise that the direction of the frictional force is opposite that of the motion, and its magnitude depends on the properties of the surfaces. Moreover, the magnitude of the frictional force also depends on how strongly the two surfaces

Charles-Augustin de Coulomb (1736-1806) was a French physicist and military engineer. For most of his working life, Coulomb served in the French army, for which he supervised many construction projects. As part of this job, Coulomb did research, first in mechanics (leading to his law of kinetic friction, equation 2.13), and later in electricity and magnetism, for which he discovered that the force between charges (and those between magnetic poles) drops off quadratically with their distance (equation 2.10). Near the end of his life, Coulomb participated in setting up the SI system of units.

Figure 2.4: Portrait of Charles de Coulomb [5].

are pushed against each other - i.e., on the forces they exert on each other, perpendicular to the surface. These forces are of course equal (by Newton's third law) and are called **normal** forces, because they are normal (that is, perpendicular) to the surface. If you stand on a box, gravity exerts a force on you pulling you down, which you 'transfer' to a force you exert on the top of the box, and causes an equal but opposite normal force exerted by the top of the box on your feet. If the box is tilted, the normal force is still perpendicular to the surface (it remains normal), but is no longer equal in magnitude to the force exerted on you by gravity. Instead, it will be equal to the component of the gravitational force along the direction perpendicular to the surface (see figure 2.6). We denote normal forces as $\boldsymbol{F}_n$. Now according to the **Coulomb friction law** (not to be confused with the Coulomb force between two charged particles), the magnitude of the frictional force between two surfaces satisfies

$$F_f \leq \mu F_n. \tag{2.13}$$

Here μ is the *coefficient of friction*, which of course depends on the two surfaces, but also on the question whether the two surfaces are moving with respect to each other or not. If they are not moving, i.e., the configuration is static, the appropriate coefficient is called the *coefficient of static friction* and denoted by μ_s. The actual magnitude of the friction force will be such that it balances the other forces (more on that in section 2.4). Equation (2.13) tells us that this is only possible if the required magnitude of the friction force is less than $\mu_s F_n$. When things start moving, the static friction coefficient is replaced by the *coefficient of kinetic friction* μ_k, which is usually smaller than μ_s; also in that case the inequality in equation (2.13) gets replaced by an equals sign, and we have $F_f = \mu_k F_n$.

2.3 EQUATIONS OF MOTION

Now that we have set our axioms - Newton's laws of motion and the various force laws - we are ready to start combining them to get useful results, things that we did not put into the axioms in the first place but follow from them. The first thing we can do is write down *equations of motion*: an equation that describes the motion of a particle due to the action of a certain type of force. For example, suppose you take a rock of a certain mass m and let go of it at some height h above the ground, then what will happen? Once you've let go of the rock, there is only one force acting on the rock, namely Earth's gravity, and we are well within the regime where equation (2.8) applies, so we know the force. We also know that this net force will result in a change of momentum (equation 2.4), which, because the rock won't loose any mass in the process of falling, can be rewritten as (2.5). By equating the forces we arrive at an equation of motion for the rock, which in this case is very simple:

$$m\boldsymbol{g} = m\ddot{\boldsymbol{x}}. \tag{2.14}$$

We immediately see that the mass of the rock doesn't matter (Galilei was right! - though of course he was in our set of axioms, because we arrived at them by assuming he was right...). Less trivially, equation (2.14) is a second-order differential equation for the motion of the rock, which means that in order to find the actual motion, we need two initial conditions - which in our present example are that the rock starts at height h and zero velocity.

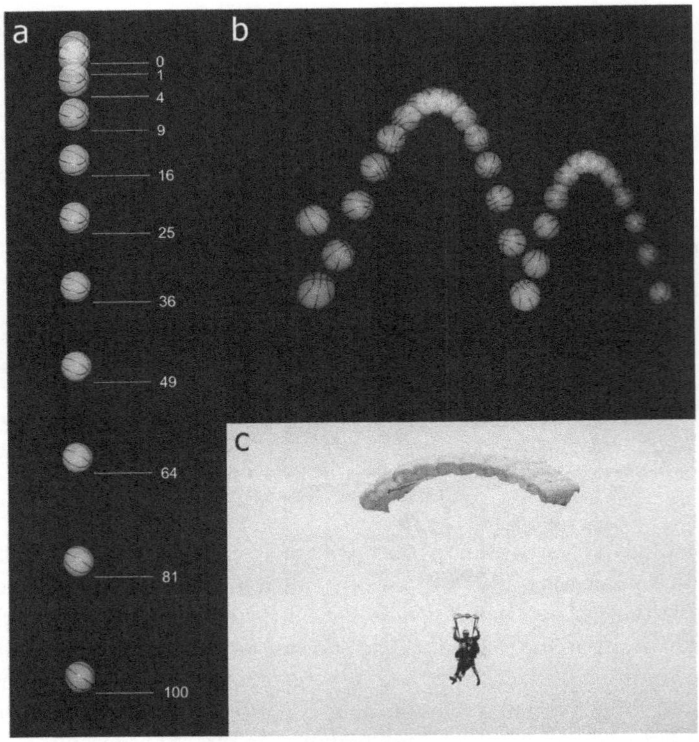

Figure 2.5: Dropping under the force of gravity. (a and b) A ball released from rest drops with a constant acceleration, resulting in a constantly increasing velocity. Images in (a) are taken every 0.05 s; distances are multiples of 12 mm. In (b), the trajectory of the ball resulting from repeated bounces is shown with intervals of 0.04 s [6], CC BY-SA 3.0. (c) Paragliders need to balance the force of gravity and that of drag to stop accelerating and fall at a continuous speed (known as their terminal velocity) [7], CC BY-SA 3.0.

Equation (2.14) is essentially one-dimensional - all motion occurs along the vertical line. Solving it is therefore straightforward - you simply integrate over time twice. The general solution is:

$$\boldsymbol{x}(t) = \boldsymbol{x}(0) + \boldsymbol{v}(0)t + \frac{1}{2}\boldsymbol{g}t^2, \tag{2.15}$$

which with our boundary conditions becomes

$$\boldsymbol{x}(t) = (h - \frac{1}{2}gt^2)\hat{\boldsymbol{z}}, \tag{2.16}$$

where g is the magnitude of $\boldsymbol{g}$ (which points down, hence the minus sign). Of course equation (2.16) breaks down when the rock hits the ground at $t = \sqrt{2h/g}$, which is easily understood because at that point gravity is no longer the only force acting on it.

We can also immediately write down the equation of motion for a mass on a spring (no gravity at present), in which the net force is given by Hooke's law. Equating that force to the net force in Newton's second law of motion gives:

$$-k\boldsymbol{x}(t) = m\ddot{\boldsymbol{x}}(t). \tag{2.17}$$

Of course, we find another second-order differential equation, so we again need the initial position and velocity to specify a solution. The general solution of (2.17) is a combination of sines and cosines, with a frequency $\omega = \sqrt{k/m}$ (as we already know from the dimensional analysis in section 1.2):

$$\boldsymbol{x}(t) = \boldsymbol{x}(0)\cos(\omega t) + \frac{\boldsymbol{v}(0)}{\omega}\sin(\omega t). \tag{2.18}$$

We'll study this case in more detail in section 8.1.

In general, the force in Newton's second law may depend on time and position, as well as on the first derivative of the position, i.e., the velocity. For the special case that it depends on only one of the three variables, we can write down the solution formally, in terms of an integral over the force. These formal solutions are given in section 2.6. To see how they work in practice, let's consider a slightly more involved problem, that of a stone falling with drag.

2.3.1 WORKED EXAMPLE: FALLING STONE WITH DRAG

Suppose we have a spherical stone of radius a that you drop from a height h at $t = 0$. At what time, and with which velocity, will the stone hit the ground? We already solved this problem in the simple case without drag above, but now let's include drag. There are then two forces acting on the stone: gravity (pointing down) with magnitude mg, and drag (pointing in the direction opposite the motion, in this case up) with magnitude $6\pi\eta a v = bv$, as given by Stokes' law (equation 2.11). Our equation of motion is now given by (with x as the height of the particle, and the downward direction as positive):

$$m\ddot{x} = -b\dot{x} + mg. \tag{2.19}$$

We see that our force does not depend on time or position, but only on velocity - so we have case 3 of appendix 2.6. We could invoke either equation (2.33) or (2.34) to write down a formal solution, but there is an easier way, which will allow us to evaluate the relevant integrals without difficulty. Since our equation of motion is linear, we know that the sum of two solutions is again a solution. One of the terms on the right hand side of equation (2.19) is constant, which means that our equation is not homogeneous (we can rewrite it to $m\ddot{x} + b\dot{x} = mg$ to see this), so a useful thing to do is to split our solution in a homogeneous and a particular part. Rewriting our equation in terms of $v = \dot{x}$ instead of x, we get $m\dot{v} + bv = mg$, from which we can immediately get a particular solution: $v_{\mathrm{p}} = mg/b$, as the time derivative of this constant v_{p} vanishes. Subtracting v_{p}, we are left with a homogeneous equation: $m\dot{v}_{\mathrm{h}} + bv_{\mathrm{h}}$, which we now solve by separation of variables. First we write $\dot{v}_{\mathrm{h}} = \mathrm{d}v_{\mathrm{h}}/\mathrm{d}t$, then re-arrange so that all factors containing v_{h} are on one side and all factors containing t are on the other, which gives $-(m/b)(1/v_{\mathrm{h}})\,\mathrm{d}v_{\mathrm{h}} = \mathrm{d}t$. We can now integrate to get:

$$-\frac{m}{b}\int_{v_0}^{v}\frac{1}{v'}\,\mathrm{d}v' = -\frac{m}{b}\log\left(\frac{v}{v_0}\right) = t - t_0, \tag{2.20}$$

which is an example of equation (2.33). After rearranging and setting $t_0 = 0$:

$$v_{\mathrm{h}}(t) = v_0 \exp\left(-\frac{b}{m}t\right). \tag{2.21}$$

Note that this homogeneous solution fits our intuition: if there is no extra force on the particle, the drag force will slow it down exponentially. Also note that we didn't set $v_0 = 0$, as the homogeneous solution does not equal the total solution. Instead v_0 is an integration constant that we'll need to set once we've written down the full solution, which is:

$$v(t) = v_{\mathrm{h}}(t) + v_{\mathrm{p}}(t) = v_0 \exp\left(-\frac{b}{m}t\right) + \frac{mg}{b}. \tag{2.22}$$

Now setting $v(0) = 0$ gives $v_0 = -mg/b$, so

$$v(t) = \frac{mg}{b}\left[1 - \exp\left(-\frac{b}{m}t\right)\right]. \tag{2.23}$$

To get $x(t)$, we simply integrate $v(t)$ over time, to get:

$$x(t) = \frac{mg}{b}\left[t + \frac{m}{b}\exp\left(-\frac{b}{m}t\right)\right]. \tag{2.24}$$

We can find when the stone hits the ground by setting $x(t) = h$ and solving for t; we can find how fast it is going at that point by substituting that value of t back into $v(t)$.

2.4 MULTIPLE FORCES

In the examples in section 2.3 there was only a single force acting on the particle of interest. Usually there will be multiple forces acting at the same time, not necessarily pulling in the same direction. This is where vectors come into play.

Suppose you put a book on a table. The Earth's gravity pulls it down with a force of magnitude F_g. Consequently the book exerts a normal force down on the table with the same magnitude, and the table reciprocates with an identical but oppositely directed normal force of magnitude $F_n = F_g$. Now suppose you push against the book from the side with a force of magnitude F. As we've seen in section 2.2, there will then be a friction force between the book and the table in the opposite direction, which, as long as it doesn't exceed $\mu_s F_n$, equals the force you push with. However, once F is larger than $\mu_s F_n$, there will be a *net force* acting on the

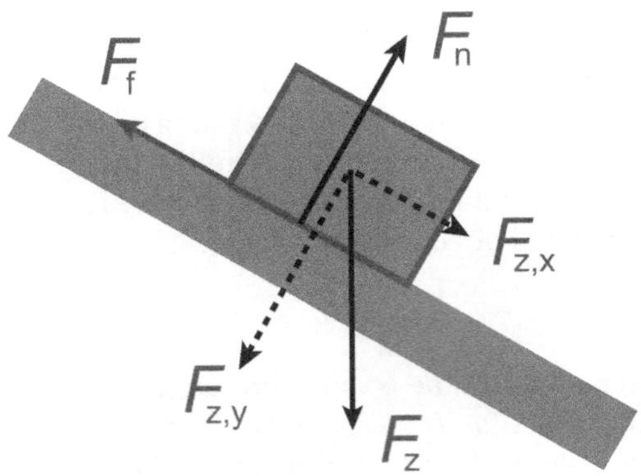

Figure 2.6: Free body diagram of the forces acting on a book on a slanted table. Gravity always points down, normal forces always perpendicular to the surface, and frictional forces always parallel to the surface. The force of gravity can be decomposed in directions perpendicular and parallel to the surface as well.

book. It is the net force that we substitute into Newton's second law, and from which the book will get a net acceleration.

In the situation described above, things are still simple - you get the net force by subtracting the kinetic friction $F_f = \mu_f F_n$ from the force F you exert on the book, because these are horizontal and thus perpendicular to the vertical normal and gravitational forces. But what happens if you lift the table on one end, so that it becomes slanted? To help organize our thoughts, we'll draw a *free body diagram*, shown in figure 2.6. Gravity still acts downward, and the mass of the book stays the same, so $\boldsymbol{F}_g$ doesn't change. However, the orientation of the contact plane between the book and table does change, so the normal force (remember, normal to the surface) changes as well. Its direction will remain perpendicular to the surface, and as long as you don't push on the book (or push along the surface only), the only other force having a component perpendicular to the surface is gravity, so the magnitude of the normal force better be equal to that (or the book would either spontaneously start to float, or fall through the table). You can find this component by decomposing the gravitational force along the directions perpendicular and parallel to the slanted surface. The remaining component of the gravitational force points downward along the surface of the table, and is comparable to the force you were exerting on the book in the flat case. Up to some point it is balanced by a static frictional force, but once it gets too large (because the slant angle of the table gets too large), friction reaches its maximum and gravity results in a net force on the book, which will start to slide down (as you no doubt guessed already).

2.5 Statics

When multiple forces act on a body, the (vector) sum of those forces gives the *net force*, which is the force we substitute in Newton's second law of motion to get the equation of motion of the body. If all forces sum up to zero, there will be no acceleration, and the body retains whatever velocity it had before. *Statics* is the study of objects that are neither currently moving nor experiencing a net force, and thus remain stationary. You might expect that this study is easier than the dynamical case when bodies do experience a net force, but that just depends on context. Imagine, for example, a jar filled with marbles: they aren't moving, but the forces acting on the marbles are certainly not zero, and also not uniformly distributed.

Even if there is no net force, there is no guarantee that an object will exhibit no motion: if the forces are distributed unevenly along an extended object, it may start to rotate. Rotations always happen around a stationary point, known as the *pivot*. Only a force that has a component perpendicular to the line connecting its point of action to the pivot (the arm) can make an object rotate. The corresponding *angular acceleration* due to the force depends on both the magnitude of that perpendicular component and the length of the arm, and is known as the *moment of the force* or the *torque* $\boldsymbol{tau}$. The magnitude of the torque is therefore given by $Fr\sin\theta$, where F is the magnitude of the force, r the length of the arm, and θ the angle between the force and the arm. If we write the arm as a vector $\boldsymbol{r}$ pointing from the pivot to the point where the force acts, we find that the magnitude of the torque equals the cross product of $\boldsymbol{r}$ and $\boldsymbol{F}$:

$$\boldsymbol{\tau} = \boldsymbol{r} \times \boldsymbol{F}. \tag{2.25}$$

The direction of rotation can be found by the right-hand rule from the direction of the torque: if the thumb of your right hand points along the direction of $\boldsymbol{\tau}$, then the direction in which your fingers curve will be the direction in which the object rotates due to the action of the corresponding force $\boldsymbol{F}$.

We will study rotations in detail in chapter 5. For now, we're interested in the case that there is no motion, neither linear nor rotational, which means that the forces and torques acting on our object must satisfy the *stability condition*: for an extended object to be stationary, both the sum of the forces and the sum of the torques acting on it must be zero.

2.5.1 WORKED EXAMPLE: SUSPENDED SIGN

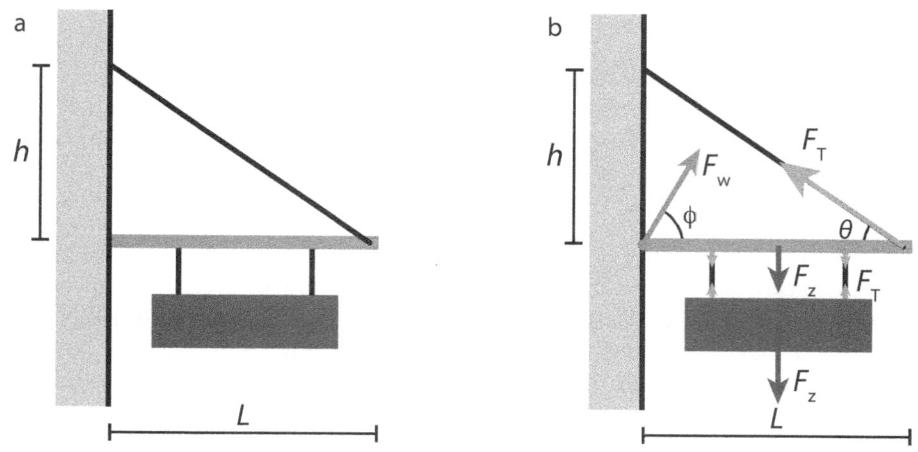

Figure 2.7: A suspended sign (example of a calculation in statics). (a) Problem setting. (b) Free-body diagram.

A sign of mass M hangs suspended from a rod of mass m and length L in a symmetric way and such that the centers of mass of the sign and rod nicely align (figure 2.7a). One end of the rod is anchored to a wall directly, while the other is supported by a wire with negligible mass that is attached to the same wall a distance h above the anchor. (a) If the maximum tension the wire can support is T, find the minimum value of h. (b) For the case that the tension in the wire equals the maximum tension, find the force (magnitude and direction) exerted by the anchor on the rod.

SOLUTION

(a) We first draw a free-body diagram, figure 2.7b. Force balance on the sign tells us that the tensions in the two lower wires sum to the gravitational force on the sign. The rod is stationary, so we know that the sum of the torques on it must vanish. To get torques, we first need a pivot; we pick the point where the rod is anchored to the wall. We then have three forces contributing a clockwise torque, and one contributing a counterclockwise torque. We're not told exactly where the wires are attached to the rod, but we are told that the configuration is symmetric and that the center of mass of the sign aligns with that of the rod. Let the first wire be a distance αL from the wall, and the second a distance $(1-\alpha)L$. The total (clockwise) torque due to the gravitational force on the sign and rod is then given by: $\tau_z = \frac{1}{2}mgL + \frac{1}{2}Mg\alpha L + \frac{1}{2}Mg(1-\alpha)L = \frac{1}{2}(m+M)gL$. The counterclockwise torque comes from the tension in the wire, and is given by $\tau_{\text{wire}} = F_T \sin\theta L = F_T(h/\sqrt{h^2+L^2})L$. Equating the two torques allows us to solve for h as a function of F_T, as requested, which gives:

$$h^2 = \left(\frac{1}{2}\frac{(m+M)g}{F_T}\right)^2 (h^2+L^2) \quad \to \quad h = \frac{(m+M)gL}{\sqrt{4F_T^2-(m+M)^2g^2}}.$$

We find the minimum value of h by substituting $F_T = T_{\max}$.

(b) As the rod is stationary, all forces on it must cancel. In the horizontal direction, we have the horizontal component of the tension, $T_{\text{max}}\cos\theta$ to the left, which must equal the horizontal component of the force exerted by the wall, $F_{\text{w}}\cos\phi$. In the vertical direction, we have the gravitational force and the two forces from the wires on which the sign hangs in the downward direction, and the vertical component of the tension in the wire in the upward direction, the sum of which must equal the vertical component of the force exerted by the wall (which may point either up or down). We thus have

$$\begin{aligned} F_{\text{w}}\cos\phi &= T_{\text{max}}\cos\theta, \\ F_{\text{w}}\sin\phi &= (m+M)g + T_{\text{max}}\sin\theta, \end{aligned}$$

where $\tan\theta = h/L$ and h is given in the answer to (a). We find that

$$\begin{aligned} F_{\text{w}}^2 &= T_{\text{max}}^2 + 2(m+M)gT_{\text{max}}\sin\theta + (m+M)^2g^2, \\ \tan\phi &= \frac{T_{\text{max}}\cos\theta}{(m+M)g + T_{\text{max}}\sin\theta}. \end{aligned}$$

Note that the above expressions give the complete answer (magnitude and direction). We could eliminate h and θ, but that'd just be algebra, leading to more complicated expressions, and not very useful in itself. If we'd been asked to calculate the height or force for any specific values of M, m, and L, we could get the answers easily by substituting the numbers in the expressions given here.

2.6 SOLVING THE EQUATIONS OF MOTION IN THREE SPECIAL CASES*

In section 2.3 we saw some examples of equations of motion originating from Newton's second law of motion. For the quite common case that the mass of our object of interest is constant, its trajectory will be given as the solution of a second-order ordinary differential equation, with time as our variable. In general, the force in Newton's second law may depend on time and position, as well as on the first derivative of the position, i.e., the velocity. In one dimension, we thus have

$$m\ddot{x} = F(x, \dot{x}, t). \tag{2.26}$$

Equation (2.26) can be hard to solve for complicated functions F. However, in each of the special cases that the force only depends on one of the three variables, we can write down a general solution - albeit as an integral over the force, which we may or may not be able to calculate explicitly.

2.6.1 CASE 1: $F = F(t)$

If the force only depends on time, we can solve equation (2.26) by direct integration. Using that $v = \dot{x}$, we have $m\dot{v} = F(t)$, which we integrate to find

$$\int_{t_0}^{t} F(t')\,\mathrm{d}t' = m\int_{v_0}^{v} \mathrm{d}v' = m[v(t) - v_0], \tag{2.27}$$

where at the initial time $t = t_0$ the object has velocity $v = v_0$. We can now find the position by integrating the velocity:

$$x(t) = \int_{t_0}^{t} v(t')\,\mathrm{d}t'. \tag{2.28}$$

2.6.2 CASE 2: $F = F(x)$

If the force depends only on the position in space (as is the case for the harmonic oscillator), we cannot integrate over time, as to do so we would already need to know $x(t)$. Instead, we invoke the chain rule to rewrite our differential equation as an equation in which the position is our variable. We have:

$$a = \frac{\mathrm{d}v}{\mathrm{d}t} = \frac{\mathrm{d}v}{\mathrm{d}x}\frac{\mathrm{d}x}{\mathrm{d}t} = v\frac{\mathrm{d}v}{\mathrm{d}x}, \tag{2.29}$$

and so our equation of motion becomes

$$mv\frac{\mathrm{d}v}{\mathrm{d}x} = F(x), \tag{2.30}$$

which we again solve by direct integration:

$$\int_{x_0}^{x} F(x')\,\mathrm{d}x' = m\int_{v_0}^{v} v'\,\mathrm{d}v' = \frac{m}{2}[v^2(x) - v_0^2]. \tag{2.31}$$

To get $x(t)$, we use the relation that $\mathrm{d}x/\mathrm{d}t = v(x)$. Separation of variables gives $\mathrm{d}x/v(x) = /ddt$, which we can integrate to get

$$t - t_0 = \int_{x_0}^{x} \frac{1}{v(x')}\,\mathrm{d}x', \tag{2.32}$$

which gives us $t(x)$. In principle we can invert this expression to give us $x(t)$, although in practice this may not be easy.

2.6.3 CASE 3: $F = F(v)$

If the force depends only on the velocity, there are two ways we can proceed. We can write the equation of motion as $m\,\mathrm{d}v/\mathrm{d}t = F(v)$ and use separation of variables to get:

$$t - t_0 = m\int_{v_0}^{v} \frac{1}{F(v')}\,\mathrm{d}v', \tag{2.33}$$

from which we can get $v(t)$ after inverting, and $x(t)$ after integrating $v(t)$ as in equation (2.28). Alternatively, we could again rewrite our equation of motion as an equation in space instead of time, and arrive at:

$$x - x_0 = m\int_{v_0}^{v} \frac{v'}{F(v')}\,\mathrm{d}v'. \tag{2.34}$$

From equation (2.34) we can get $v(x)$ by inverting, and $x(t)$ from equation (2.32). Note that equation (2.34) does not give us $x(t)$ directly, as x is the variable in that equation.

2.6.4 WORKED EXAMPLE: VELOCITY OF THE HARMONIC OSCILLATOR

It may seem that what we've done so far in this section has hardly helped matters: the 'solutions' we found contain integrals and often need to be inverted to get our desired function $x(t)$ (or, depending on the problem we're studying, $v(t)$ or $v(x)$). To show you how these solutions may be useful, let's consider a specific example: a harmonic oscillator, consisting of a mass on a Hookean spring, with $F = F(x) = -kx$. We already wrote down the equation of motion (2.17) and its general solution (2.18). The general solution can be found through the substitution of exponentials, as we'll do in section 8.1. However, we can also learn something useful from writing the equation of motion in the form (2.30). Its solution, formally given by equation (2.31), can be calculated explicitly for our force as

$$\frac{m}{2}[v^2(x) - v_0^2] = \int_{x_0}^{x} (-kx')\,\mathrm{d}x' = -\frac{k}{2}[x^2 - x_0^2], \tag{2.35}$$

which gives

$$v(x) = \sqrt{v_0^2 - \frac{k}{m}\left(x^2 - x_0^2\right)} \tag{2.36}$$

for $v(x)$. Although $x(t)$ and $v(t)$ are more easily obtained from the solution given in equation (2.18), that solution will not give you $v(x)$, and deriving it is tricky. Here we get it almost for free. Moreover, as you have probably noted, equation (2.35) relates the kinetic to the potential energy of the harmonic oscillator - a special case of conservation of energy, which we'll discuss in the next section.

2.7 PROBLEMS

2.1 The *terminal velocity* is the maximum (constant) velocity a dropping object reaches. In this problem, we use equation (2.12) for the drag force.

(a) Use dimensional analysis to relate the terminal velocity of a falling object to the various relevant parameters.

(b) Estimate the terminal velocity of a paraglider (figure 2.5c).

(c) Use the concept of terminal velocity to predict whether a mouse (without parachute) is likely to survive a fall from a high tower.

2.2 When you cook rice, some of the dry grains always stick to the measuring cup. A common way to get them out is to turn the measuring cup upside-down and hit the bottom (now on top) with your hand so that the grains come off [32].

(a) Explain why static friction is irrelevant here.

(b) Explain why gravity is negligible.

(c) Explain why hitting the cup works, and why its success depends on hitting the cup hard enough.

2.3 A ball is thrown at speed v from zero height on level ground. We want to find the angle θ at which it should be thrown so that the area under the trajectory is maximized.

(a) Sketch of the trajectory of the ball.

(b) Use dimensional analysis to relate the area to the initial speed v and the gravitational acceleration g.

(c) Write down the x and y coordinates of the ball as a function of time.

(d) Find the total time the ball is in the air.

(e) The area under the trajectory is given by $A = \int y \, \mathrm{d}x$. Make a variable transformation to express this integral as an integration over time.

(f) Evaluate the integral. Your answer should be a function of the initial speed v and angle θ.

(g) From your answer at (f), find the angle that maximizes the area, and the value of that maximum area. Check that your answer is consistent with your answer at (b).

2.4 If a mass m is attached to a given spring, its period of oscillation is T. If two such springs are connected end to end, and the same mass m is attached, find the new period T', in terms of the old period T.

2.5 Two blocks, of mass m and $2m$, are connected by a massless string and slide down an inclined plane at angle θ. The coefficient of kinetic friction between the lighter block and the plane is μ, and that between the heavier block and the plane is 2μ. The lighter block leads.

(a) Find the magnitude of the acceleration of the blocks.

(b) Find the tension in the taut string.

2.6 A 1000 kg boat is traveling at 100 km/h when its engine is shut off. The magnitude F_d of the drag force between the boat and the water is proportional to the speed v of the boat, with a drag coefficient $\zeta = 70\,\mathrm{N \cdot s/m}$. Find the time it takes the boat to slow to 45 km/h.

2.7 Two particles on a line are mutually attracted by a force $F = -ar$, where a is a constant and r the distance of separation. At time $t = 0$, particle A of mass m is located at the origin, and particle B of mass $m/4$ is located at $r = 5.0$ cm.

(a) If the particles are at rest at $t = 0$, at what value of r do they collide?

(b) What is the relative velocity of the two particles at the moment the collision occurs?

2.8 In drag racing, specially designed cars maximize the friction with the road to achieve maximum acceleration. Consider a drag racer (or 'dragster') as shown in figure 2.8, for which the center of mass is close to the rear wheels.

Figure 2.8: A drag racer or dragster [8], CC BY-SA 3.0.

(a) Draw a free-body diagram of the dragster in side view. Draw the wheels as circles, and approximate the shape of the dragster body as a triangle with a horizontal line between the wheels, a vertical line going up from the rear axis, and a diagonal line connecting the top to the front wheels. NB: consider carefully the direction of the friction force!

(b) On which of the wheels is the frictional force the largest?

(c) The frictional force is maximized if the wheels just don't slip (because, as usual, the coefficient of kinetic friction is smaller than that of static friction). Find the maximal possible frictional force on the rear wheels.

(d) Find the maximal possible acceleration of the dragster.

(e) For a coefficient of (static) friction of 1.0 (a fairly realistic value for rubber and concrete) and a track of 500 m, find the maximal velocity a drag racer can achieve at the end of the track when starting from rest.

2.9 Blocks A, B and C are placed as shown in the figure, and connected by ropes of negligible mass. Both A and B weigh 20.0 N each, and the coefficient of kinetic friction between each block and the surface is 0.3. The slope's angle θ equals 42.0°. The disks in the pulleys are of negligible mass. After the blocks are released, block C descends with constant velocity.

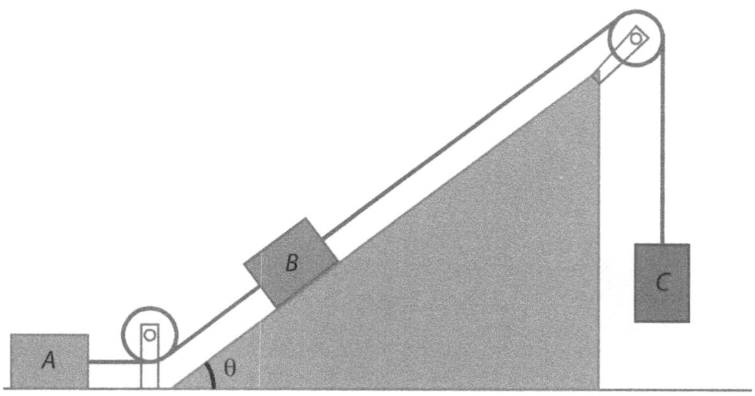

(a) Find the tension in the rope connecting blocks A and B.

(b) What is the weight of block C?

(c) If the rope connecting blocks A and B were cut, what would be the acceleration of C?

2.10 The figure below shows a common present-day seesaw design. In addition to a beam with two seats, this seesaw also contains two identical springs that connect the beam to the ground. The distance between the pivot and each of the springs is 30.0 cm, the distance between the pivot and each of the seats is 1.50 m.

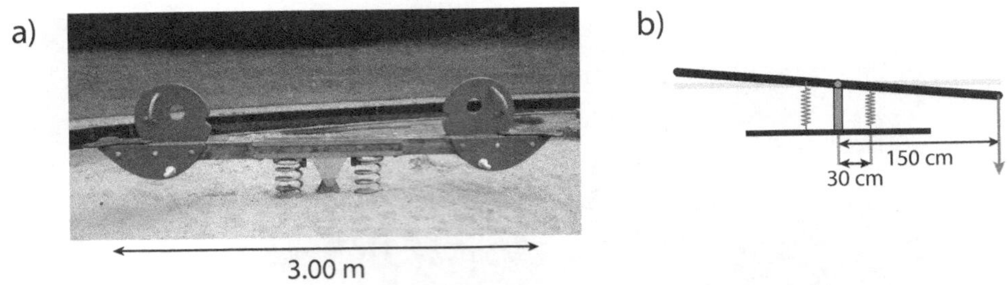

(a) A 4-year-old weighing 20.0 kg sits on one of the seats, causing it to drop by 20.0 cm. Draw a free-body diagram of the seesaw with the child, in which you include all relevant forces (to scale).

(b) Use your diagram and the provided data to calculate the spring constant of the two springs present in the seesaw.

2.11 Two marbles of identical mass m and radius r are dropped in a cylindrical container with radius $3r$, as shown in the figure. Find the force exerted by the marbles on points A, B and C, and the force the marbles exert on each other.

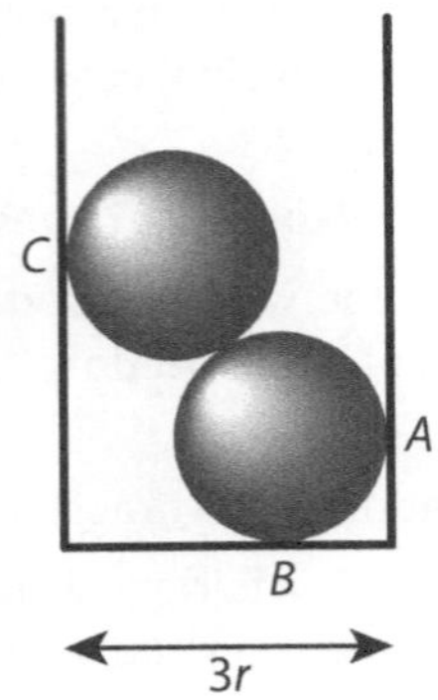

2.12 Round fruits like oranges and mandarins are typically stacked in alternating rows, as shown in figure 2.9. Suppose you have a crate with a square base that is exactly five oranges wide. You stack 25 oranges in the crate, then put another 16 on top in the holes, and then add a second layer of 25 oranges, held in place by the sides of the crate. Find the total force on the sides of the crate in this configuration. Assume all oranges are spheres with a diameter of 8.0 cm and a mass of 250 g.

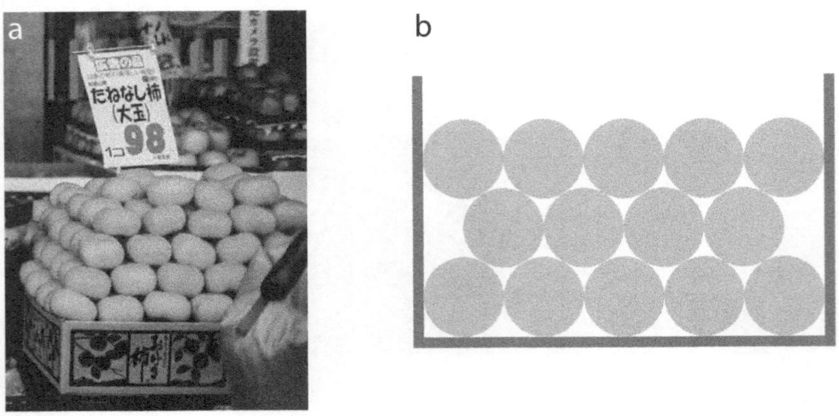

Figure 2.9: Stacked fruit. (a) Stacked mandarins at a fruit stand [9]. (b) Cross-section of stacked oranges in a crate.

2.13 Objects with densities less than that of water float, and even objects that have higher densities are 'lighter' in the water. The force that's responsible for this is known as the *buoyancy force*, which is equal but opposite to the gravitational force on the displaced water: $F_{\text{buoyancy}} = \rho_{\text{w}} g V_{\text{w}}$, where ρ_{w} is the water's density and V_{w} the displaced volume. In parts (a) and (b), we consider a block of wood with density $\rho < \rho_{\text{w}}$ which is floating in water.

(a) Which fraction of the block of wood is submerged when floating?

(b) You push down the block somewhat more by hand, then let go. The block then oscillates on the surface of the water. Explain why, and calculate the frequency of the oscillation.

(c) You take out the piece of wood, and now float a piece of ice in a bucket of water. On top of the ice, you place a small stone. When everything has stopped moving, you mark the water level. Then you wait till the ice has melted, and the stone has dropped to the bottom of the bucket. What has happened to the water level? Explain your answer (you can do so either qualitatively through an argument or quantitatively through a calculation).

(d) Rubber ducks also float, but, despite the fact that they have a flat bottom, they usually do not stay upright in water. Explain why.

(e) You drop a 5.0 kg ball with a radius of 10 cm and a drag coefficient c_d of 0.20 in water (viscosity 1.002 mPa·s). This ball has a density higher than that of water, so it sinks. After a while, it reaches a constant velocity, known as its terminal velocity. What is the value of this terminal velocity?

(f) When the ball in (e) has reached terminal velocity, what is the value of its Reynolds number (see problem 1.3)?

2.14 A uniform stick of mass M and length $L = 1.00$ m has a weight of mass m hanging from one end. The stick and the weight hang in balance on a force scale at a point $x = 20.0$ cm from the end of the stick. The measured force equals 3.00 N. Find both the mass M of the stick and m of the weight.

2.15 A uniform rod with a length of 4.25 m and a mass of 47.0 kg is attached to a wall with a hinge at one end. The rod is held in horizontal position by a wire attached to its other end. The wire makes an angle of 30.0° with the horizontal, and is bolted to the wall directly above the hinge. If the wire can support a maximum tension of 1250 N before breaking, how far from the wall can a 75.0 kg person sit without breaking the wire?

2.16 A wooden bar of uniform density but varying thickness hangs suspended on two strings of negligible mass. The strings make angles θ_1 and θ_2 with the horizontal, as shown. The bar has total mass m and length L. Find the distance x between the center of mass of the bar and its (thickest) end.

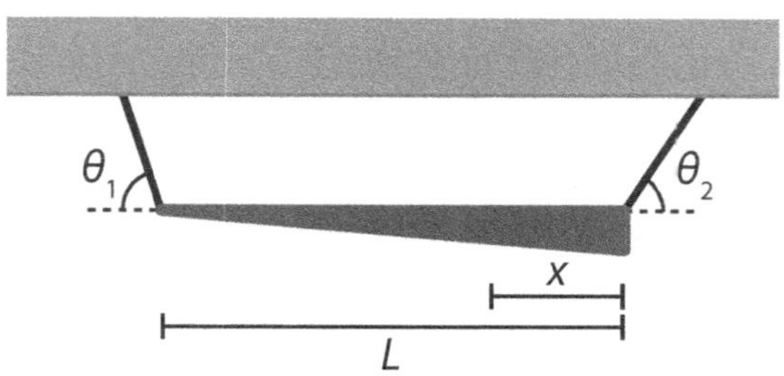

2.17 A bicycle wheel of radius R and mass M is at rest against a step of height $3R/4$, as shown in the figure.Find the minimum horizontal force F that must be applied to the axle to make the wheel start to rise over the step.

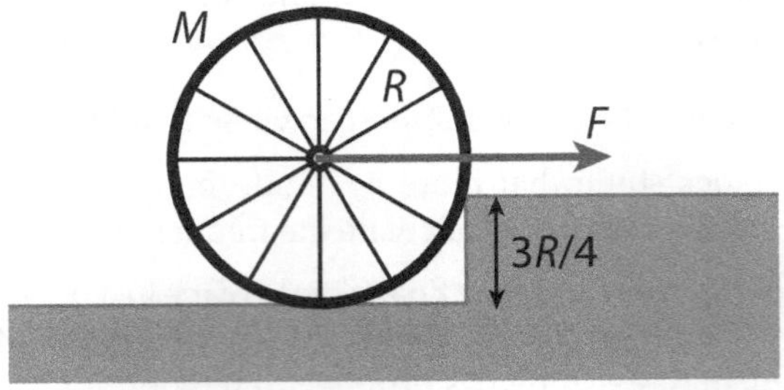

2.18 A block of mass M is pressed against a vertical wall, with a force F applied at an angle θ with respect to the horizontal ($-\pi/2 < \theta < \pi/2$), as shown in the figure. The friction coefficient of the block and the wall is μ. We start with the case $\theta = 0$, i.e., the force is perpendicular to the wall.

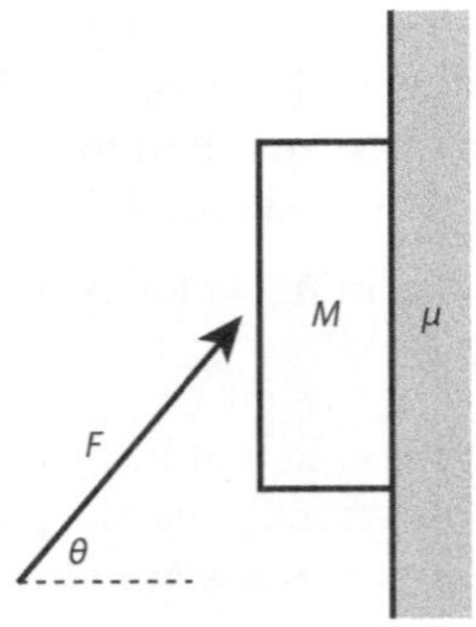

(a) Draw a free-body diagram showing all forces.
(b) If the block is to remain stationary, the net force on it should be zero. Write down the equations for force balance (i.e., the sum of all forces is zero, or forces in one direction equal the forces in the opposite direction) for the x and y directions.
(c) From the two equations you found in (b), solve for the force F needed to keep the book in place.
(d) Now repeat the steps you took in (a)-(c) for a force under a given angle θ, and find the required force F.
(e) For what angle θ is this minimum force F the smallest? What is the corresponding minimum value of F?
(f) What is the limiting value of θ, below which it is not possible to keep the block up (independent of the magnitude of the force)?

2.19 A spherical stone of mass $m = 0.250$ kg and radius $R = 5.0$ cm is launched vertically from ground level with an initial speed of $v_0 = 15.0$ m/s. As it moves upwards, it experiences drag from the air as approximated by Stokes drag, $F = 6\pi\eta R v$, where the viscosity η of air is 1.002 mPa·s.

(a) Which forces are acting on the stone while it moves upward?
(b) Using Newton's second law of motion, write down an equation of motion for the stone (this is a differential equation). Be careful with the signs. *Hint*: Newton's second law of motion relates force and acceleration, and the drag force is in terms of the velocity. What is the relation between the two? Simplify the equation by introducing the characteristic time $\tau = \frac{m}{6\pi\eta R}$.
(c) Find a particular solution $v_\mathrm{p}(t)$ of your inhomogeneous differential equation from (19b).
(d) Find the solution $v_\mathrm{h}(t)$ of the homogeneous version of your differential equation.
(e) Use the results from (19c) and (19d) and the initial condition to find the general solution $v(t)$ of your differential equation.
(f) From (19e), find the time at which the stone reaches its maximum height.
(g) From $v(t)$, find $h(t)$ for the stone (height as a function of time).
(h) Using your answers to (19f) and (19g), find the maximum height the stone reaches.

3

ENERGY

3.1 WORK

How much work do you need to do to move a box? Well, that depends on two things: how heavy the box is, and how far you have to move it. Multiply the two, and you've got a good measure of how much work will be required. Of course, work can be done in other contexts as well - pulling a spring from equilibrium, or cycling against the wind. In each case, there's a *force* and a *displacement*. To be fair, we will only count the part of the force that is in the direction of the displacement (when cycling, you don't do work due to the fact that there's a gravitational force pulling you down, since you don't move vertically; you do work because there's a drag force due to your moving through the air). We define *work* as the product of the component of the force in the direction of the displacement, times the displacement itself. We calculate this component by projecting the force vector on the displacement vector, using the dot product (see appendix A.1 for an introduction in to vector math):

$$W = \boldsymbol{F} \cdot \boldsymbol{x}. \tag{3.1}$$

Note that work is a scalar quantity - it has a magnitude but no direction. Work is measured in Joules (J), with one Joule being equal to one Newton times one meter.

Of course the force acting on our object need not be constant everywhere. Take for example the extension of a spring: the further you pull, the larger the force gets, as given by Hooke's law (2.7). To calculate the work done when extending the spring, we chop up the path (here a straight line) into many small pieces. For each piece, we approximate the force by the average value on that piece, then multiply with the length of the piece and sum. In the limit that we have infinitely many pieces, this approximation becomes exact, and the sum becomes an integral: for one dimension, we thus have:

$$W = \int_{x_1}^{x_2} F(x)\,\mathrm{d}x. \tag{3.2}$$

Likewise, the path along which we move need not be a straight line. If the path consists of multiple straight segments, on each of which the force is constant, we can calculate the total work by adding the work done on the different segments. Taking the limit to infinitely many infinitesimally small segments $\mathrm{d}\boldsymbol{r}$, on each of which the force is given by the value $\boldsymbol{F}(\boldsymbol{r})$, the sum again becomes an integral:

$$W = \int_{\boldsymbol{r}_1}^{\boldsymbol{r}_2} \boldsymbol{F}(\boldsymbol{r}) \cdot \mathrm{d}\boldsymbol{r}. \tag{3.3}$$

Equation (3.3) is the most general version of the definition of work; it simplifies to (3.2) for movement along a straight line, and to (3.1) if both the path is straight and the force constant[1].

[1] If you feel intimidated by the vector form of equation (3.3), it may help to rewrite it in terms of the magnitudes of the force $\boldsymbol{F}(\boldsymbol{r})$ and the (infinitesimal) displacement $\mathrm{d}\boldsymbol{r}$, and the angle θ between them. In terms of $F = |\boldsymbol{F}|$, $\mathrm{d}r = |\mathrm{d}\boldsymbol{r}$ and θ, we have $\boldsymbol{F} \cdot \mathrm{d}\boldsymbol{r} = F \cos\theta \,\mathrm{d}r$, an expression you may have seen before for a force not pointing in the same direction as the displacement. If we now make the force and displacements functions of the position $\boldsymbol{r}$, then so become the magnitude of the force and the angle, so we can also write equation (3.3) as

$$W = \int_{\boldsymbol{r}_1}^{\boldsymbol{r}_2} F(\boldsymbol{r}) \cos\theta(\boldsymbol{r}) \cdot \mathrm{d}r. \tag{3.4}$$

In general, the work done depends on the path taken - for example, it's more work to take a detour when biking from home to work, assuming the air drag is the same everywhere. However, in many important cases the work done in getting from one point to another depends on the endpoints only. Forces for which this is true are called *conservative forces*. As we'll see below, the force exerted by a spring and that exerted by gravity are both conservative.

Sometimes we will not be interested in how much work is done in generating a certain displacement, but over a certain amount of time - for instance, a generator generates work by getting something to move, like a wheel or a valve, but we don't typically care about those details, we want to know how much work we can expect to get out of the generator, i.e., how much *power* it has. Power is defined as the amount of work per unit time, or

$$P = \frac{\mathrm{d}W}{\mathrm{d}t}. \tag{3.5}$$

Power is measured in Joules per second, or Watts (W). To find out how much work is done by an engine that has a certain power output, we need to integrate that output over time:

$$W = \int P \,\mathrm{d}t. \tag{3.6}$$

3.2 KINETIC ENERGY

Newton's first law told us that a moving object will stay moving unless a force is acting on it - which holds for moving with any speed, including zero. Now if you want to start moving something that is initially at rest, you'll need to accelerate it, and Newton's second law tells you that this requires a force - and moving something means that you're displacing it. Therefore, there is work involved in getting something moving. We define the *kinetic energy* (K) of a moving object to be equal to the work required to bring the object from rest to that speed, or equivalently, from that speed to rest:

$$K = \frac{1}{2} m v^2. \tag{3.7}$$

Because the kinetic energy is equal to an amount of work, it is also a scalar quantity, has the same dimension, and is measured in the same unit. The factor v^2 is the square of the magnitude of the velocity of the moving object, which you can calculate with the dot product: $v^2 = \boldsymbol{v} \cdot \boldsymbol{v}$. You may wonder where equation (3.7) comes from. Newton's second law tells us that $\boldsymbol{F} = m \,\mathrm{d}\boldsymbol{v}/\mathrm{d}t$, relating the force to an infinitesimal change in the velocity. In the definition for work, equation (3.3), we multiply the force with an infinitesimal change in the position $\mathrm{d}\boldsymbol{r}$. That infinitesimal displacement takes an infinitesimal amount of time $\mathrm{d}t$, which is related to the displacement by the instantaneous velocity $\boldsymbol{v}$: $\mathrm{d}\boldsymbol{r} = \boldsymbol{v}\,\mathrm{d}t$. We can now calculate the work necessary to accelerate from zero to a finite speed:

$$K = \int \boldsymbol{F} \cdot \mathrm{d}\boldsymbol{r} = \int m \frac{\mathrm{d}\boldsymbol{v}}{\mathrm{d}t} \cdot \boldsymbol{v} \,\mathrm{d}t = \int m\boldsymbol{v} \cdot \frac{\mathrm{d}\boldsymbol{v}}{\mathrm{d}t} \,\mathrm{d}t = \int m\boldsymbol{v} \cdot \mathrm{d}\boldsymbol{v} = \frac{m}{2} \int \mathrm{d}(\boldsymbol{v} \cdot \boldsymbol{v}) = \frac{1}{2} m v^2, \tag{3.8}$$

where we used that the dot product is commutative and the fact that the integral over the derivative of a function is the function itself.

Of course, now that we know that the kinetic energy is given by equation (3.7), we no longer need to use a complicated integral to calculate it. However, because the kinetic energy is ultimately given by this integral, which is equal to a net amount of work, we arrive at the following statement, sometimes referred to as the **Work-energy theorem**: the change in kinetic energy of a system equals the net amount of work done on or by it (in case of increase/decrease of K):

$$\Delta K = W_{\mathrm{net}}. \tag{3.9}$$

Figure 3.1: Examples of high power resulting in high kinetic energy. (a) Running cheetah, the fastest land animal, which can reach speeds over 100 km/h in 2-3 seconds, corresponding to an enormous increase in its kinetic energy [10], CC BY-SA 3.0. (b) Allyson Felix running second in the women's 4 × 400 relay of the 2012 London Summer Olympics [11], CC BY-SA 3.0. (c) Robert Garrett preparing to throw the discus at the 1896 Athens Summer Olympics [12]. Unlike the runners, the goal of discus throwing is to maximize the distance, not the speed, but to get the largest possible distance, the discus must still get the maximal possible kinetic energy.

3.3 POTENTIAL ENERGY

We already encountered *conservative forces* in section 3.1. The work done by a conservative force is (by definition) path-independent; that means that in particular the work done when moving along any closed path[2] must be zero:

$$\oint \boldsymbol{F} \cdot \mathrm{d}\boldsymbol{r} = 0. \tag{3.10}$$

For a conservative force, we can thus define a *potential energy difference* between points 1 and 2 as the work necessary to move an object from point 1 to point 2:

$$\Delta U_{12} = -\int_{\boldsymbol{r}_1}^{\boldsymbol{r}_2} \boldsymbol{F} \cdot \mathrm{d}\boldsymbol{r}. \tag{3.11}$$

Note the minus sign in the definition - this is a choice of course, and you'll see below why we made this choice. Note also that the potential energy is defined only between two points. Often we will choose a convenient reference point and calculate the potential energy at any other point with respect to that point. The reference point is typically either the origin or infinity, if the force happens to be zero at either of these. Let's suppose we have set such a point, and know the potential energy difference with that point at any other point in space - this defines a (scalar) function $U(\boldsymbol{r})$. If we now want to know the force acting on a particle at $\boldsymbol{r}$, all we need to do is take the derivative of $U(\boldsymbol{r})$ - that is to say the gradient in three dimensions (which simplifies to the ordinary derivative in one dimension):

$$\boldsymbol{F}(\boldsymbol{r}) = -\boldsymbol{\nabla} U(\boldsymbol{r}). \tag{3.12}$$

Equation (3.12) is extremely useful, as it gives us a means to calculate the force, which is a vector quantity, from the potential energy function, which is a scalar quantity - and therefore much simpler to work with. For instance, since energies are scalars, they can simply be added, as we'll do in the next section, whereas for forces you need to do vector addition. Equation (3.12) also reflects that we are free to choose a reference point for the potential energy, since the force does not change if we add a constant to the potential energy.

3.3.1 GRAVITATIONAL POTENTIAL ENERGY

We saw in section 2.2.2 that for low altitudes, the gravitational force is given by $\boldsymbol{F}_g = m\boldsymbol{g}$, where $\boldsymbol{g}$ is a vector of constant magnitude $g \approx 9.81\mathrm{m/s}^2$ and always points down. Therefore, the gravitational force does no work when you move horizontally, and if you first move up and then the same amount down again, it doesn't do any net work either, as the two contributions exactly cancel. $\boldsymbol{F}_g$ is therefore an example of a conservative force, and we can define and calculate the *gravitational potential energy* U_g between a point at height 0 (our reference point) and one at height h:

$$U_g(h) = -\int_{z=0}^{z=h} m(-g)\,\mathrm{d}z = mgh. \tag{3.13}$$

Note that by choosing a minus sign in the definition of the potential energy, we end up with a positive value of the energy here.

What about larger distances, i.e., Newton's law of gravity, equation (2.9)? Well, there the distances are measured radially, so any movement perpendicular to the radial direction doesn't matter, and if you move out and back in again, the net work done is zero, so by the same reasoning as before we again have a conservative force. This force vanishes at infinity, so it makes sense to set that as a reference point - though notice that that will make our potential energy always negative in this case:

$$U_G(r) = -\frac{GMm}{r} \tag{3.14}$$

where r is the distance between m and M, and M sits at the origin. Of course we can also calculate gravitational potential differences between two distances r_1 and r_2 from M: $\Delta U_G(r_1, r_2) = GMm\left(\frac{1}{r_1} - \frac{1}{r_2}\right)$.

[2]The integral sign with the circle in equation (3.10) represents an integral over a closed path.

Emmy Noether (1882-1935) was a German mathematician, who made key contribution both to the development of abstract algebra and to ideas in theoretical physics. In physics, she uncovered a deep connection between symmetry and conservation laws (now known as *Noether's theorem*, considered by many as the most important theorem for the development of modern physics): for every continuous symmetry of a system, there exists a conserved quantity. A continuous symmetry is one that leaves a system invariant for an arbitrarily large given transformation; for example, the rotation of a circle under any angle. Applications of Noether's theorem include conservation of energy (corresponding to invariance under time translation, i.e., it doesn't matter where you set $t = 0$, section 3.4), conservation of momentum (invariance under space translation, i.e., it doesn't matter where you put the origin, section 4.2) and conservation of angular momentum (invariance under space rotation, i.e., it doesn't matter in which direction you choose your x-axis, section 5.7). Similar conservation laws are found in special and general relativity, quantum mechanics, and quantum field theory. Unfortunately, even in the early 20th century, women were still excluded from most academic positions. Noether therefore initially worked for free at the university of Erlangen, getting a paid position in Göttingen in 1915 at the invitation of Hilbert and Klein, who had both been convinced by the quality of her work. Her fame grew through the 1910s and 1920s, gaining worldwide recognition. Due to her Jewish descent, she was dismissed from her academic position by the Nazi government in 1933, and moved to the United States, where she died two years later at age 53. Various institutes and scholarship programs, mostly in Germany, are now named in her honor.

Figure 3.2: Emmy Noether [13].

3.3.2 SPRING POTENTIAL ENERGY

Like the gravitational force, the Hookean spring force (2.7) also depends on displacement alone, and by the same reasoning is conservative (notice the pattern?). Calculating its associated potential energy is straightforward, and taking the equilibrium position of the spring as the reference point, we find:

$$U_s(x) = \frac{1}{2}kx^2. \tag{3.15}$$

The minus sign in Hooke's Law gives us a positive spring potential energy. Note that x stands for displacement here; as we only consider one-dimensional springs the 1D-version is sufficient.

3.3.3 GENERAL CONSERVATIVE FORCES

In the case of the gravitational and spring force it was easy to reason that they had to be conservative. It is also easy to see that the friction force is not conservative: if you take a longer path, you need to do more net work against friction, which you can moreover never recover as mechanical energy. For more complicated systems, especially in three dimensions, it may not be so easy to see whether a force is conservative. Fortunately, there is an easy test you can perform: if the curl of a force is zero everywhere, it will be a conservative force, or expressed mathematically:

$$\nabla \times \boldsymbol{F} = 0 \quad \Leftrightarrow \quad \oint \boldsymbol{F} \cdot \mathrm{d}\boldsymbol{r} = 0 \quad \Leftrightarrow \quad \boldsymbol{F} = -\nabla U. \tag{3.16}$$

Is is straightforward to show that if a force is conservative, its curl must vanish: a conservative force can be written as the gradient of some scalar function $U(\boldsymbol{x})$, and $\nabla \times \nabla U(\boldsymbol{x}) = 0$ for any function $U(\boldsymbol{x})$, as you can easily check for yourself. The proof the other way around is more complicated, and can be found in advanced mechanics textbooks.

3.4 CONSERVATION OF ENERGY

Work, kinetic energy and potential energy are all quantities with the same dimension - so we can do arithmetic with them. One particularly useful quantity is the *total energy* E of a system, which is simply the sum of the kinetic and potential energy:

$$E = K + U. \tag{3.17}$$

Theorem 3.1 (Law of conservation of energy). *If all forces in a system are conservative, the total energy in that system is conserved.*

Proof. For simplicity, we'll look at the 1D case (3D goes analogously). Conserved means not changing in time, so in order to prove the statement, we only need to calculate the time derivative of E and check that it is always zero.

$$\begin{aligned}\frac{\mathrm{d}E}{\mathrm{d}t} &= \frac{\mathrm{d}K}{\mathrm{d}t} + \frac{\mathrm{d}U}{\mathrm{d}t}\\ &= \frac{\mathrm{d}\left(\frac{1}{2}mv^2\right)}{\mathrm{d}t} + \frac{\mathrm{d}U}{\mathrm{d}x}\frac{\mathrm{d}x}{\mathrm{d}t}\\ &= mv\frac{\mathrm{d}v}{\mathrm{d}t} - Fv\\ &= -\left(F - m\frac{\mathrm{d}v}{\mathrm{d}t}\right)v\\ &= 0 \text{ by Newton's second law.}\end{aligned} \tag{3.18}$$

□

Conservation of energy means that the total energy of a system cannot change, but of course the potential and kinetic energy can - and by conservation of total energy we know that they get converted directly into one another. Exploiting this fact will allow us to analyze and easily solve many problems in classical mechanics - this conservation law is an immensely useful tool.

Note that conservation of energy is not the same as the work-energy theorem of section 3.2. For the total energy to be conserved, all forces need to be conservative. In the work-energy theorem, this is not the case. You can therefore calculate changes in kinetic energy due to the work done by non-conservative forces using the latter.

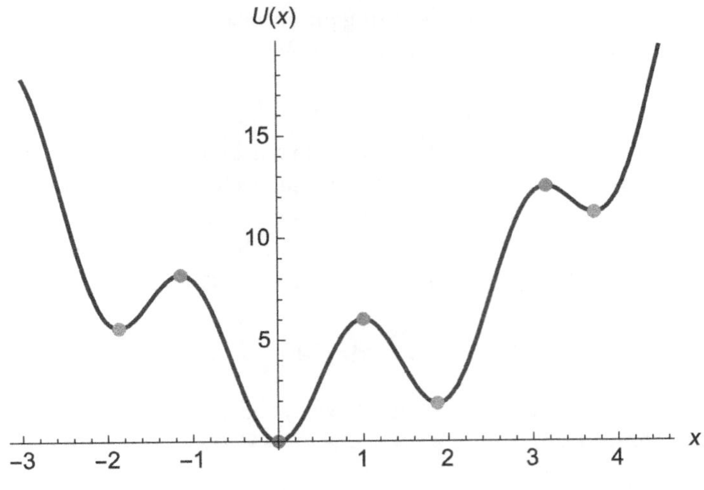

Figure 3.3: An example of a potential energy landscape. In this figure, the total energy would be represented by a horizontal line; the kinetic energy by the distance between the potential and total energy. Equlibrium points (dots) occur at extrema of the potential energy, when its derivative (the force) is zero. The green dots indicate unstable equilibrium points (maxima, where the second derivative is negative), the orange points metastable equilibria (local minima) and the red point the single globally stable equilibrium of this system.

3.5 ENERGY LANDSCAPES

In the previous section we proved that the total energy is conserved. In the section before that, we looked at potential energies. Typically, the potential energy is a function of your position in space. When we plot it as a function of spatial coordinates, we get an *energy landscape*, measuring an amount of energy on the vertical axis. Of course we can also plot the total energy of the system - and since that is conserved, it is the same everywhere, and thus becomes a horizontal line or plane. Because kinetic energy cannot be negative, any point where the potential energy is higher than the total energy is not allowed: the system cannot reach this point. When the potential energy equals the total energy, the kinetic energy (and thus the speed) has to be zero. Whenever the potential energy is lower than the total energy, there is a positive kinetic energy and thus a positive speed.

Probably the simplest energy landscape is that of the harmonic oscillator (mass on a spring) - it's a simple parabola. The point at which the horizontal line representing the total energy crosses the parabola corresponds to the extrema of the oscillation: these are its turning points. The bottom of the parabola is its midpoint, and you can immediately see that that's where the kinetic energy (and thus the speed) will be highest.

Of course you can have more complex energy landscapes than that. In particular, you can have a landscape with multiple extrema, see for example figure 3.3. A particle that is being acted upon by forces described by this potential energy, follows a trajectory in this landscape, which can be visualized as a ball rolling over the hills and valleys of the landscape. Think back to the harmonic oscillator example. If we let go of a ball in a parabolic vase at some point on the slope, the ball will roll down and pick up speed, then roll up the opposite slope and loose speed, until it reaches the same height where its speed will again be zero. The same is true in more complicated landscapes. Particularly interesting are local maxima. If you put a ball exactly on top of one of them, it will stay there - it is a fixed point, but an unstable one, as any arbitrarily small perturbation will push it down. If you let go of a ball at a level above a local maximum, it may hop over it to the next minimum, but if your initial position (your initial energy) was too low, your ball can get stuck oscillating about a local minimum - a metastable point.

Energy landscapes are even useful when the total energy is not conserved - for example because of friction terms. Friction causes energy to dissipate from the system, which is equivalent to having your ball move in the landscape with friction. For low friction, your ball will oscillate, but get less high every time, until it comes to rest at the minimum. For high friction, it won't even oscillate, but just get to the minimum - exactly what an overdamped system in real life does.

3.5.1 WORKED EXAMPLE: THE LENNARD-JONES POTENTIAL

The Lennard-Jones potential energy is a commonly used model to describe the interactions between uncharged atoms and molecules. This potential energy can be written in two equivalent ways:

$$U_{\mathrm{LJ}}(r) = \frac{A}{r^{12}} - \frac{B}{r^6} = 4\varepsilon\left[\left(\frac{\sigma}{r}\right)^{12} - \left(\frac{\sigma}{r}\right)^6\right], \tag{3.19}$$

where r is the distance between the atoms or molecules, and A, B, ε and σ are positive constants.

(a) Find the dimensions of A, B, ε and σ.

(b) Express ε and σ in A and B.

(c) Sketch the potential (in its second form) as a function of r/σ, and use this sketch to give a physical interpretation of ε and σ.

(d) Does the Lennard-Jones potential lead to attractive or repulsive forces at short distances? And what about long distances?

(e) Find all equilibrium points of this potential energy, and determine their stability.

SOLUTION

(a) $[U] = \text{Energy} \Longrightarrow [U] = M \times \frac{L}{T^2} \times L = \frac{ML^2}{T^2}$
$[A] = \text{Energy} \times \text{Length}^{12} \Longrightarrow [A] = \frac{ML^{14}}{T^2}$
$[B] = \text{Energy} \times \text{Length}^6 \Longrightarrow [A] = \frac{ML^8}{T^2}$
Because the powers of the terms $((\frac{\sigma}{r})^{12})$ and $((\frac{\sigma}{r})^6)$ are different, while we add them together, they have to be dimensionless, so $[\sigma] = L$ and $[\varepsilon] = [U] = \frac{ML^2}{T^2}$.

(b)

$$4\varepsilon\sigma^{12} = A \text{ and } 4\varepsilon\sigma^6 = B$$

$$\frac{A}{B} = \sigma^6 \Longrightarrow \sigma = \left(\frac{A}{B}\right)^{1/6}.$$

By substituting σ in the expressions for either A or B we can derive an expression for ε:

$$4\sigma^6\varepsilon = B$$

$$4\varepsilon A = B^2 \Longrightarrow \varepsilon = \frac{B^2}{4A}.$$

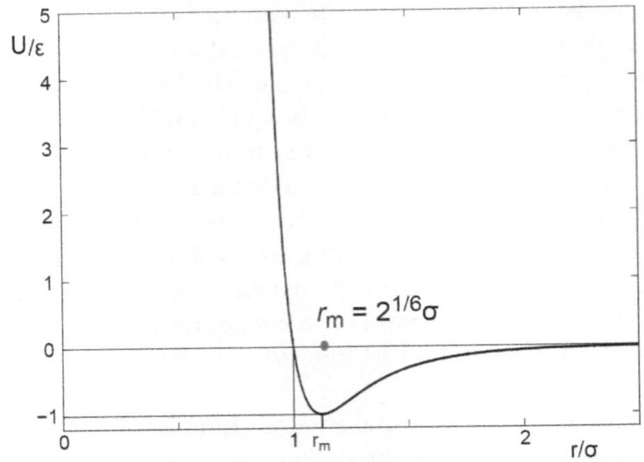

Figure 3.4: Sketch of the Lennard-Jones potential energy.

(c) See figure 3.4. Interpretation: ε is a measure for the depth of the potential well. σ sets the length scale and therefore the position of the equilibrium point.

(d) *Method 1*: We calculate the force as minus the derivative of the potential energy:

$$F = -\frac{\partial U}{\partial r} = 4\varepsilon\left(\frac{12\sigma^{12}}{r^{13}} - \frac{6\sigma^6}{r^7}\right)$$

For small r we have $r^{-13} >> r^{-7}$, so F is positive and therefore repulsive. Conversely, for large r we have $r^{-13} << r^{-7}$, so F is negative and therefore attractive.
Method 2: Use the sketch in (c) to to see that the slope of the potential is negative for small r, which implies a repulsive force, and the slope of the potential is positive for large r, which implies an attractive force.

(e) For an equilibrium point we have:

$$0 = \frac{\partial U}{\partial r} = 4\varepsilon\left(\frac{12\sigma^{12}}{r^{13}} - \frac{6\sigma^6}{r^7}\right) = 24\frac{\varepsilon\sigma^6}{r^7}\left(\frac{2\sigma^6}{r^6} - 1\right), \tag{3.20}$$

so there is only one equilibrium point, at

$$r_{\mathrm{eq}} = 2^{1/6}\sigma. \tag{3.21}$$

To determine the stability at this point, we consider the second derivative of $U(r)$:

$$\left.\frac{\partial^2 U}{\partial r^2}\right|_{r=r_{\mathrm{eq}}} = 4\varepsilon\left.\left(42\frac{\sigma^6}{r^8} - 156\frac{\sigma^{12}}{r^{14}}\right)\right|_{r=r_{\mathrm{eq}}} = 4\varepsilon\left(\frac{42}{2^{4/3}\sigma^2} - \frac{156}{2^{7/3}\sigma^2}\right) = -36\cdot 2^{2/3}\frac{\varepsilon}{\sigma^2} < 0, \tag{3.22}$$

which means that the equilibrium point is stable. Alternatively, we could have determined the stability by considering the graph drawn at (c), from which we can see that the equilibrium point corresponds to a global minimum of the potential energy and hence is stable.

3.6 PROBLEMS

3.1 (a) Show that, if you ignore drag, a projectile fired at an initial velocity v_0 and angle θ has a range R given by

$$R = \frac{v_0^2 \sin 2\theta}{g}. \tag{3.23}$$

(b) A target is situated 1.5 km away from a cannon across a flat field. Will the target be hit if the firing angle is 42° and the cannonball is fired at an initial velocity of 121 m/s? (Cannonballs, as you know, do not bounce).

(c) To increase the cannon's range, you put it on a tower of height h_0. Find the maximum range in this case, as a function of the firing angle and velocity, assuming the land around is still flat.

3.2 You push a box of mass m up a slope with angle θ and kinetic friction coefficient μ. Find the minimum initial speed v you must give the box so that it reaches a height h.

3.3 A uniform board of length L and mass M lies near a boundary that separates two regions. In region 1, the coefficient of kinetic friction between the board and the surface is μ_1, and in region 2, the coefficient is μ_2. Our objective is to find the net work W done by friction in pulling the board directly from region 1 to region 2, under the assumption that the board moves at constant velocity.

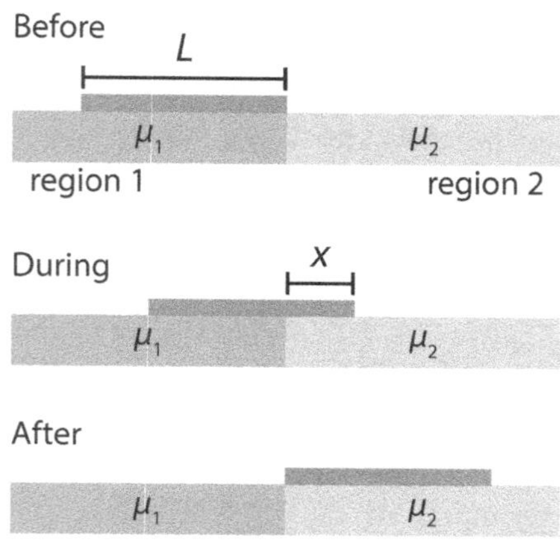

(a) Suppose that at some point during the process, the right edge of the board is a distance x from the boundary, as shown. When the board is at this position, what is the magnitude of the force of friction acting on the board, assuming that it's moving to the right? Express your answer in terms of all relevant variables (L, M, g, x, μ_1, and μ_2).

(b) As we've seen in section 3.1, when the force is not constant, you can determine the work by integrating the force over the displacement, $W = \int F(x)\,dx$. Integrate your answer from (a) to get the net work you need to do to pull the board from region 1 to region 2.

3.4 The government wishes to secure votes from car-owners by increasing the speed limit on the highway from 120 to 140 km/h. The opposition points out that this is both more dangerous and will cause more pollution. Lobbyists from the car industry tell the government not to worry: the drag coefficients of the cars have gone down significantly and their construction is a lot more solid than in the time that the 120 km/h speed limit was set.

(a) Suppose the 120 km/h limit was set with a Volkswagen Beetle ($c_d = 0.48$) in mind, and the lobbyist's car has a drag coefficient of 0.19. Will the new car need to do more or less work to maintain a constant speed of 140 km/h than the Beetle at 120 km/h?

(b) What is the ratio of the total kinetic energy released in a full head-on collision (resulting in an immediate standstill) between two cars both at 140 km/h and two cars both at 120 km/h?

(c) The government dismisses the opposition's objections on safety by stating that on the highway, all cars move in the same direction (opposite direction lanes are well separated), so if they all move at 140 km/h, it would be just as safe as all at 120 km/h. The opposition then points out that running a

Beetle (those are still around) at 120 km/h is already challenging, so there would be speed differences between newer and older cars. The government claims that the 20 km/h difference won't matter, as clearly even a Beetle can survive a 20 km/h collision. Explain why their argument is invalid.

3.5 Nuclear fusion, the process that powers the Sun, occurs when two low-mass atomic nuclei fuse together to make a larger nucleus, releasing substantial energy. Fusion is hard to achieve because atomic nuclei carry positive electric charge, and their electrical repulsion makes it difficult to get them close enough for the short-range nuclear force to bind them into a single nucleus. The figure below shows the potential-energy curve for fusion of two deuterons (heavy hydrogen nuclei, consisting of a proton and a neutron). The energy is measured in million electron volts (MeV, 1 eV = $1.6 \cdot 10^{-19}$ J), a unit commonly used in nuclear physics, and the separation is in femtometers (1 fm = 10^{-15} m).

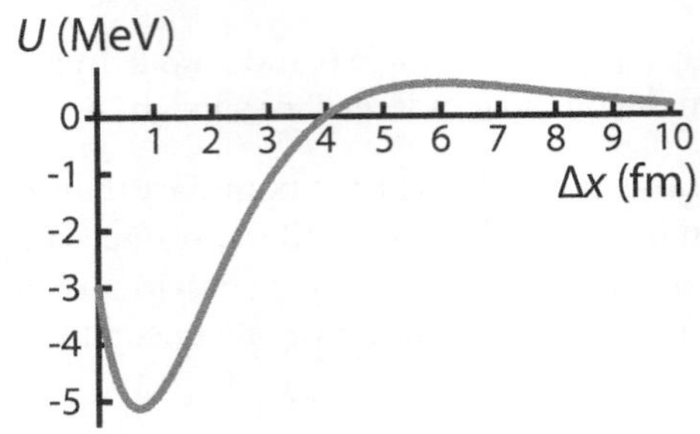

(a) Find the position(s) (if any) at which the force between two deuterons is zero.

(b) Find the kinetic energy two initially widely separated deuterons need to have to get close enough to fuse.

(c) The energy available in fusion is the energy difference between that of widely separated deuterons and the bound deutrons after they've 'fallen' into the deep potential well shown in the figure. About how big is that energy?

(d) Determine whether the force between two deuterons that are 4 fm apart is repulsive, attractive, or zero.

3.6 A pigeon in flight experiences a drag force due to air resistance given approximately by $F = bv^2$, where v is the flight speed and b is a constant.

(a) What are the units of b?

(b) What is the largest possible speed of the pigeon if its maximum power output is P?

(c) By what factor does the largest possible speed increase if the maximum power output is doubled?

3.7 (a) For which value(s) of the parameters α, β and γ is the force given by

$$\boldsymbol{F} = (x^3y^3 + \alpha z^2, \beta x^4 y^2, \gamma xz)$$

conservative?

(b) Find the force for the potential energy given by $U(x, y, z) = xy/z - xz/y$.

3.8 A point mass is connected to two opposite walls by two springs, as shown in the figure. The distance between the walls is $2L$. The left spring has rest length $l_1 = L/2$ and spring constant $k_1 = k$, the right spring has rest length $l_2 = 3L/4$ and spring constant $k_2 = 3k$.

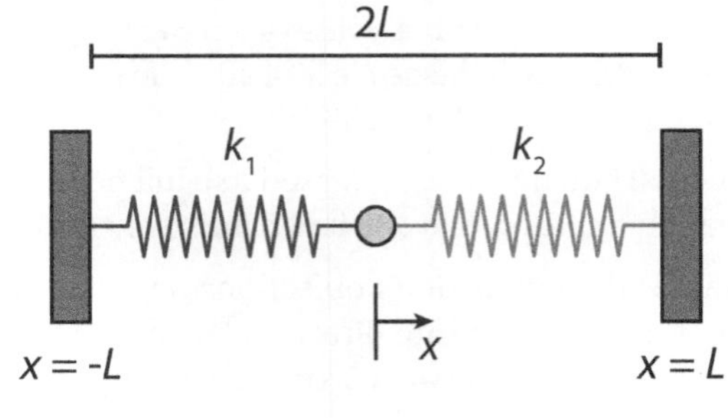

(a) Determine the magnitude of the force acting on the point mass if it is at $x = 0$.

(b) Determine the equilibrium position of the point mass.

(c) Find the potential energy of the point mass as a function of x. Use the equilibrium point from (b) as your point of reference.

(d) If the point mass is displaced a small distance from its equilibrium position and then released, it will oscillate. By comparing the equation of the net force on the mass in this system with a simple harmonic oscillator, determine the frequency of that oscillation. (We'll return to systems oscillating about the minimum of a potential energy in section 8.1.4, feel free to take a sneak peak ahead).

3.9 A block of mass $m = 3.50\ kg$ slides from rest a distance d down a frictionless incline at angle $\theta = 30.0°$, where it runs into a spring of spring constant 450 N/m. When the block momentarily stops, it has compressed the spring by 25.0 cm.

(a) Find d.

(b) What is the distance between the first block-spring contact and the point at which the block's speed is greatest?

3.10 Playground slides frequently have sections of varying slope: steeper ones to pick up speed, less steep ones to loose speed, so kids (and students) arrive at the bottom safely. We consider a slide with two steep sections (angle α) and two less steep ones (angle β). Each of the sections has a width L. The slide has a coefficient of kinetic friction μ.

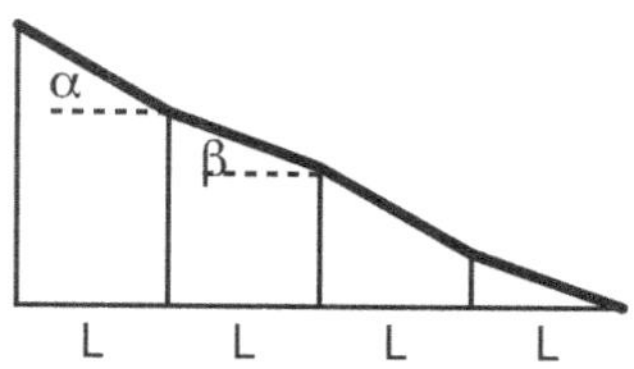

(a) Kids start at the top of the slide with velocity zero. Calculate the velocity of a kid of mass m at the end of the first steep section.

(b) Now calculate the velocity of the kid at the bottom of the entire slide.

(c) If $L = 1.0$ m, $\alpha = 30°$ and $\mu = 0.5$, find the minimum value β must have so that kids up to 30 kg can enjoy the slide (*Hint*: what is the minimum requirement for the slide to be functional)?

(d) A given slide has $\alpha = 30°$, $\beta = 25°$ and $\mu = 0.5$. A young child of 10 kg slides down, and continues sliding on the ground after reaching the bottom of the slide. The coefficient of kinetic friction with the ground is 0.70. How far does the child slide before it comes to a full stop?

3.11 In this problem, we consider the anharmonic potential given by

$$U(x) = \frac{a}{2}(x - x_0)^2 + \frac{b}{3}(x - x_0)^3, \tag{3.24}$$

where a, b and x_0 are positive constants.

(a) Find the dimensions of a, b and x_0.

(b) Determine whether the force on a particle at a position $x >> x_0$ is attractive or repulsive (taking the origin as your point of reference).

(c) Find the equilibrium point(s) (if any) of this potential, and determine their stability.

(d) For $b = 0$, the potential given in equation (3.24) becomes harmonic (i.e., the potential of a harmonic oscillator), in which case a particle that is initially located at a non-equilibrium point will oscillate. Are there initial values for x for which a particle in this anharmonic potential will oscillate? If so, find them, *and* find the approximate oscillation frequency; if not, explain why not. (NB: As the problem involves a third order polynomial function, you may find yourself having to solve a third order problem. When that happens, for your answer you can simply say: the solution x to the problem X).

3.12 After you have successfully finished your mechanics course, you decide to launch the book into an orbit around the Earth. However, the teacher is not convinced that you do not need it anymore and asks the following question: What is the ratio between the kinetic energy and the potential energy of the book in its orbit?

Let m be the mass of the book, $M_\oplus$ and $R_\oplus$ the mass and the radius of the Earth respectively. The gravitational pull at distance r from the center is given by Newton's law of gravitation (equation 2.9):

$$\boldsymbol{F}_g(r) = -G\frac{mM_\oplus}{r^2}\hat{\boldsymbol{r}}$$

(a) Find the orbital velocity v of an object at height h above the surface of the Earth.

(b) Express the work required to get the book at height h.

(c) Calculate the ratio between the kinetic and the potential energy of the book in its orbit.

(d) What requires more work, getting the book to the International Space Station (orbiting at $h = 400$ km) or giving it the same speed as the ISS?

3.13 Using dimensional arguments, in problem 1.4 we found the scaling relation of the escape velocity (the minimal initial velocity an object must have to escape the gravitational pull of the planet/moon/other object it's on completely) with the mass of the radius of the planet. Here, we'll re-derive the result, including the numerical factor that dimensional arguments cannot give us.

(a) Derive the expression of the gravitational potential energy, U_g, of an object of mass m due to a gravitational force F_g given by Newton's law of gravitation (equation 2.9):

$$\boldsymbol{F}_g = -\frac{GmM}{r^2}\hat{\boldsymbol{r}}.$$

Set the value of the integration constant by $U_g \to 0$ as $r \to \infty$.

(b) Find the escape velocity on the surface of a planet of mass M and radius R by equating the initial kinetic energy of your object (when launched from the surface of the planet) to the total gravitational potential energy it has there.

3.14 A cannonball is fired upwards from the surface of the Earth with just enough speed such that it reaches the Moon. Find the speed of the cannonball as it crashes on the Moon's surface, taking the gravity of both the Earth and the Moon into account. Table B.3 contains the necessary astronomical data.

3.15 The draw force $F(x)$ of a Turkish bow as a function of the bowstring displacement x (for $x < 0$) is approximately given by a quadrant of the ellipse

$$\left(\frac{F(x)}{F_{\max}}\right)^2 + \left(\frac{x+d}{d}\right)^2 = 1.$$

In rest, the bowstring is at $x = 0$; when pulled all the way back, it's at $x = -d$.

(a) Calculate the work done by the bow in accelerating an arrow of mass $m = 37$ g, for $d = 0.85$ m and $F_{\max} = 360$ N.

(b) Assuming that all of the work is converted to kinetic energy of the arrow, find the maximum distance the arrow can fly. *Hint*: which variable can you control when shooting? Maximize the distance with respect to that variable.

(c) Compare the result of (b) with the range of a bow that acts like a simple (Hookean) spring with the same values of $F_{\max}$ and d. How much further does the arrow shot from the Turkish bow fly than that of the simple spring bow?

3.16 A massive cylinder with mass M and radius R is connected to a wall by a spring at its center (see figure). The cylinder can roll back-and-forth without slipping.

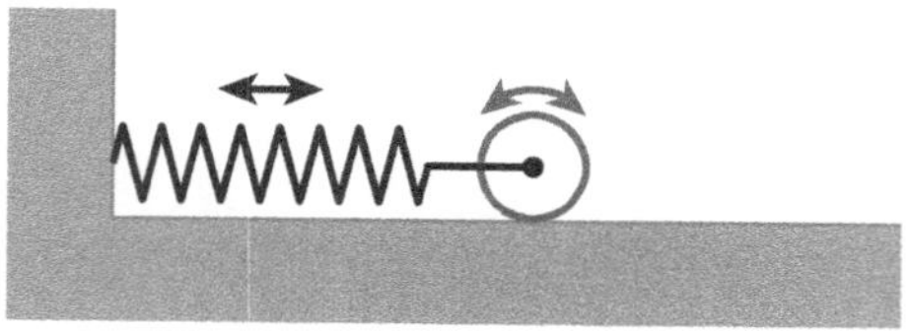

(a) Determine the total energy of the system consisting of the cylinder and the spring.

(b) Differentiate the energy of problem (16a) to obtain the equation of motion of the cylinder and spring system.

(c) Find the oscillation frequency of the cylinder by comparing the equation of motion at (16b) with that of a simple harmonic oscillator (a mass-spring system).

3.17 A small particle (blue dot) is placed atop the center of a hemispherical mount of ice of radius R (see figure). It slides down the side of the mount with negligible initial speed. Assuming no friction between the ice and the particle, find the height at which the particle loses contact with the ice.

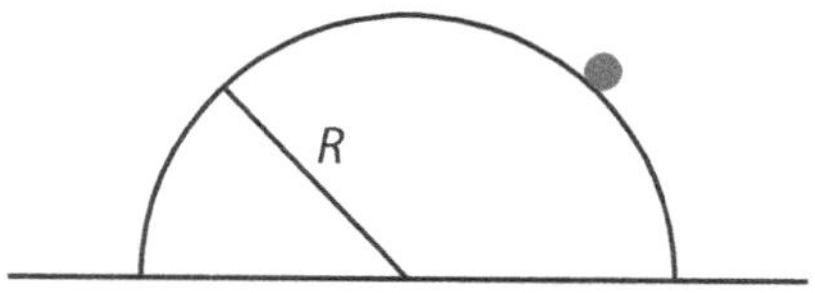

Hint: To solve this problem, first draw a free body diagram, and combine what you know of energy and forces.

3.18 **Pulling membrane tubes**

The (potential) energy of a cylindrical membrane tube of length L and radius R is given by

$$\mathscr{E}_{\text{tube}}(R,L) = 2\pi RL\left(\frac{\kappa}{2}\frac{1}{R^2}+\sigma\right). \tag{3.25}$$

Here κ is the membrane's bending modulus and σ its surface tension.

(a) Find the dimensions of the bending modulus and the surface tension.

(b) Find the forces acting on the tube along its radial and axial direction.

(c) Membrane tubes are often pulled by membrane motors pulling along the axial direction, as sketched in figure 3.5. For that case, we add the work done by the motors to the total energy of the tube, so we get:

$$\mathscr{E}_{\text{tube}}(R,L) = 2\pi RL\left(\frac{\kappa}{2}\frac{1}{R^2}+\sigma\right) - FL. \tag{3.26}$$

Show that for a stable tube, the motors need to exert a force of magnitude $F = 2\pi\sqrt{2\kappa\sigma}$.

(d) Can the force of (c) be considered to be an effective spring force? If so, find its associated spring constant. If not, explain why not.

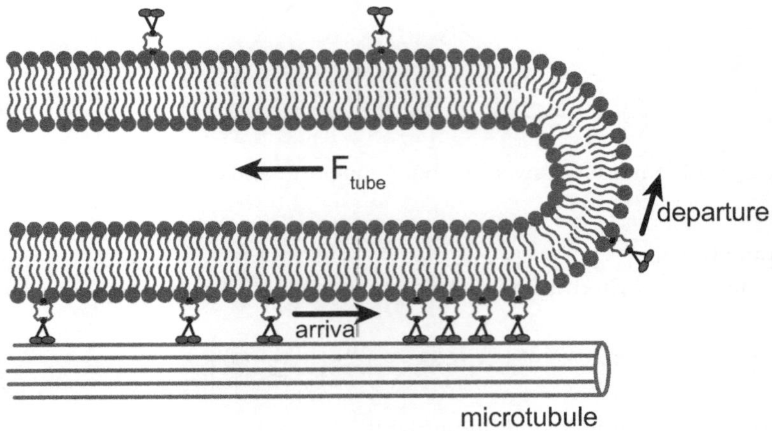

Figure 3.5: Cartoon of molecular motors together pulling a membrane tube.

4

MOMENTUM

4.1 CENTER OF MASS

4.1.1 CENTER OF MASS OF A COLLECTION OF PARTICLES

So far we've only considered two cases - single particles on which a force is acting (like a mass on a spring), and pairs of particles exerting a force on each other (like gravity). What happens if more particles enter the game? Well, then we have to calculate the total force, by vector addition, and total energy, by regular addition. Let's label the particles with a number α, then the total force is given by:

$$\boldsymbol{F}_{\text{total}} = \sum_\alpha \boldsymbol{F}_\alpha = \sum_\alpha m_\alpha \ddot{\boldsymbol{r}}_\alpha = M\frac{\mathrm{d}^2}{\mathrm{d}t^2}\left(\frac{\sum_\alpha m_\alpha \boldsymbol{r}_\alpha}{M}\right) = M\frac{\mathrm{d}^2}{\mathrm{d}t^2}\boldsymbol{r}_{\text{cm}}, \tag{4.1}$$

where we've defined the *total mass* $M = \sum_\alpha m_\alpha$ and the *center of mass*

$$\boldsymbol{r}_{\text{cm}} = \frac{1}{M}\sum_\alpha m_\alpha \boldsymbol{r}_\alpha. \tag{4.2}$$

4.1.2 CENTER OF MASS OF AN OBJECT

Equation (4.2) gives the center of mass of a discrete set of particles. Of course, in the end, every object is built out of a discrete set of particles, its molecules, but summing them all is going to be a lot of work. Let's try to do better. Consider a small sub-unit of the object of volume $\mathrm{d}V$ (much smaller than the object, but much bigger than a molecule). Then the mass of that sub-unit is $\mathrm{d}m = \rho\,\mathrm{d}V$, where ρ is the density (mass per unit volume) of the object. Summation over all these masses gives us the center of mass of the object, by equation (4.2). Now taking the limit that the volume of the sub-units goes to zero, this becomes an infinite sum over infinitesimal volumes - an integral. So for the center of mass of a continuous object we find:

$$\boldsymbol{r}_{\text{cm}} = \frac{1}{M}\int_V \rho\cdot\boldsymbol{r}\,\mathrm{d}V. \tag{4.3}$$

Note that in principle we do not even need to assume that the density ρ is constant - if it depends on the position in space, we can also absorb that in the discussion above, and end up with the same equation, but now with $\rho(\boldsymbol{r})$. That will make the integral a lot harder to evaluate, but not necessarily impossible. Also note that the total mass M of the object is simply given by $\rho\cdot V$, where V is the total volume, if the density is constant, and by $\int_V \rho(\boldsymbol{r})\,\mathrm{d}V$ otherwise. Therefore, if the density is constant, it drops out of equation (4.3), and we can rewrite it as

$$\boldsymbol{r}_{\text{cm}} = \frac{1}{V}\int_V \boldsymbol{r}\,\mathrm{d}V \qquad \text{for constant density } \rho. \tag{4.4}$$

Unfortunately, many textbooks introduce the confusing concept of a infinitesimal mass element $\mathrm{d}m$, instead of a volume element $\mathrm{d}V$ with mass $\rho\,\mathrm{d}V$. This strange habit often throws students off, and the concept is wholly unnecessary, so we won't adapt it here.

Equation (4.3) holds for any continuous object, but it might be confusing if you consider a linear or planar object - as you may wonder how the density ρ and volume element $\mathrm{d}V$ are defined in one and two dimensions. There are two ways out. One is to say that all physical objects are three-dimensional - even a very thin stick has a cross section. If you say that cross section has area A (which is constant along the stick, or the thin stick approximation would be invalid), and the coordinate along the stick is x, the volume element simply becomes $\mathrm{d}V = A\,\mathrm{d}x$, and the integral in equation (4.3) reduces to a one-dimensional integral. You can approach two-dimensional objects in the same way, by giving them a small thickness Δz and writing the volume element as $\mathrm{d}V = \Delta z\,\mathrm{d}A$. Alternatively, you can define one- and two-dimensional analogs of the density: the mass per unit length λ and mass per unit area σ, respectively. With those, the one- and two-dimensional equivalents of equation (4.3) are given by

$$x_{\mathrm{cm}} = \frac{1}{M}\int_0^L \lambda x\,\mathrm{d}x, \text{ and } \boldsymbol{r}_{\mathrm{cm}} = \frac{1}{M}\int_A \rho \cdot \boldsymbol{r}\,\mathrm{d}A,$$

where M is still the total mass of the object.

4.1.3 Worked example: center of mass of a solid hemisphere

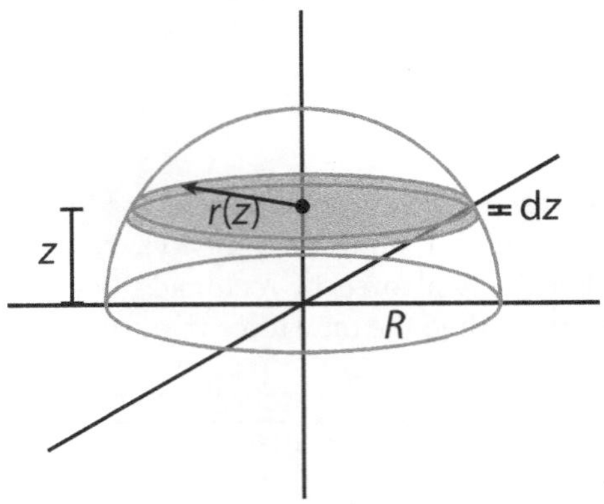

Figure 4.1: Coordinate system for the calculation of the center of mass for a solid hemisphere.

By symmetry, the center of mass of a solid sphere must lie at its center. The center of mass of a hemisphere cannot be guessed so easily, so we must calculate it. Of course, it must still lie on the axis of symmetry, but to calculate where on that axis, we'll use equation 4.4. To carry out the integral, we'll make use of the symmetry the system still has, and chop our hemisphere up into thin slices of equal thickness $\mathrm{d}z$, see figure 4.1. The volume of such a slice will then depend on its position z, and be given by $\mathrm{d}V = \pi r(z)^2\,\mathrm{d}z$, where $r(z)$ is the radius at height z. Putting the origin at the bottom of the hemisphere, we easily obtain $r(z) = \sqrt{R^2 - z^2}$, where R is the radius of the hemisphere. The position vector $\boldsymbol{r}$ in equation 4.4 simply becomes $(0,0,z)$, so we get:

$$z_{\mathrm{cm}} = \frac{1}{\frac{2}{3}\pi R^3}\int_0^R z\pi(R^2 - z^2)\,\mathrm{d}z = \frac{3}{2R^3}\left[\frac{1}{2}z^2R^2 - \frac{1}{4}z^4\right]_0^R = \frac{3}{8}R. \tag{4.5}$$

The center of mass of the solid hemisphere thus lies at $\boldsymbol{r}_{\mathrm{cm}} = (0,0,3R/8)$.

4.2 Conservation of momentum

In equation (4.1), what is the total force acting on all the particles? Well, that's the sum of all the forces the particles exert on each other, plus all external forces: $\boldsymbol{F}_{\mathrm{total}} = \sum_i \boldsymbol{F}_{\mathrm{int},i} + \sum_i \boldsymbol{F}_{\mathrm{ext},i}$. Now Newton's third law of motion tells us that the internal forces come in opposite pairs, so when we sum them, they all cancel, and the total net force acting on the particles is equal to the sum of the external forces acting on the particles. Therefore, by equation (4.1), *the center of mass of a system of particles obeys Newton's second law of motion.*

What about the momentum of the center of mass? Like the force, the total momentum of the system of the system is given by the vector sum of the individual particle momenta:

$$\boldsymbol{P}_{\mathrm{total}} = \sum_\alpha \boldsymbol{p}_\alpha = \sum_\alpha m_\alpha \dot{\boldsymbol{r}}_\alpha = \frac{\mathrm{d}}{\mathrm{d}t}\sum_\alpha m_\alpha \boldsymbol{r}_\alpha = \frac{\mathrm{d}}{\mathrm{d}t}M\boldsymbol{r}_{\mathrm{cm}}, \tag{4.6}$$

so the total momentum of the system equals that of the center of mass. Moreover, as long as the mass of the system is conserved, we can rewrite equation (4.1) as

$$F_{\text{total}} = \frac{dP_{\text{total}}}{dt}. \tag{4.7}$$

Not only does the center of mass of a system of particles obey Newton's second law of motion, its total momentum does too. Moreover, unlike in the single-particle case, equation (4.7) has an important consequence for the case that there is *no* external force acting on the system. For one particle, that would simply mean that the momentum does not change - Newton's first law of motion. But for multiple particles, equation (4.7) tells us that no external forces means that the *total* momentum does not change. We have therefore arrived at our second conservation law:

Theorem 4.1 (Law of conservation of momentum). *When no external forces act on a system of particles, the total momentum of the system is conserved.*

We derived the law of conservation of momentum by applying both Newton's second and third laws of motion, so like conservation of energy, it is not an independent result, but follows from our axioms. Note that the law allows for the momenta of the individual particles in the system to change, as long as their total stays the same - this is what happens when you play billiards, and why the number of balls bouncing in a Newton's cradle is fixed.

4.3 REFERENCE FRAMES

4.3.1 CENTER OF MASS FRAME

The center of mass need not be fixed in space, so it can have a nonzero velocity, which of course is simply given by $v_{\text{cm}} = \dot{r}_{\text{cm}}$. For each of the particles in a multi-particle system, we can *decompose* its velocity by writing it as the sum of the center of mass velocity and a velocity relative to the center of mass:

$$v_\alpha = v_{\text{cm}} + v_{\alpha,\text{rel}}. \tag{4.8}$$

In many applications, the information is in the velocity component relative to the center of mass. After all, conservation of momentum implies that for a system with no external forces acting on it, the center of mass velocity cannot change, even if all the individual momenta do change (as happens in collisions). Therefore, it is often convenient to analyze your system in a frame that moves with the center of mass, known (unsurprisingly), as the center of mass frame. In this frame, the center of mass velocity is identically zero, and again because of conservation of momentum, all other velocities in this frame must sum to zero. The 'real-world' frame with nonzero center of mass velocity is referred to as the lab frame.

4.3.2 GALILEAN TRANSFORMATIONS AND INERTIAL FRAMES

As equation (4.8) shows, if you know a particle's velocity in the center of mass frame, you can easily calculate the velocity in the lab frame by adding the velocity of the center of mass. Going the other way, to calculate the velocity in the center of mass frame, you subtract v_{cm} from the velocity in the lab frame. Moreover, if the center of mass moves at constant velocity, we can also easily relate positions in both frames. If we denote coordinates in the lab frame by r and those in the center of mass frame by r', we readily obtain:

$$r = r' + v_{\text{cm}}\, t, \tag{4.9}$$

$$v = v' + v_{\text{cm}}. \tag{4.10}$$

Equation (4.9) is an example of a *Galilean transformation* between frames of reference (here the lab frame and the center of mass frame). It actually holds for any pair of reference frames that move with constant velocity with respect to each other. Such frames of reference are known as *inertial frames* if Newton's first law of motion holds in them; by Newton's second law, if one of the frames is an inertial frame, then the one obtained from it by a Galilean transformation (i.e., one moving at constant velocity with respect to the first frame) is also an inertial frame. The reason for this is that a constant velocity plays no role in Newton's second law, as it relates the derivative of the velocity (i.e., the acceleration) to a force. Consequently, not only is Newton's first law of motion valid in both inertial frames - all laws of physics are the same in two such frames. This fact is known as the *principle of relativity*. It does not apply to, for example, frames that rotate with respect to each other, as we'll see in chapter 7. Moreover, although the principle of relativity is universally valid (it is in fact one of the two basic assumptions behind Einstein's theory of relativity), the Galilean transformations are not. They break down at velocities that approach the speed of light, as we'll explore in detail in part II.

4.3.3 KINETIC ENERGY OF A COLLECTION OF PARTICLES

We've established above that both the total momentum and energy are conserved in closed systems, but the components can of course change. Momentum can be transferred from one particle to another, and so can (kinetic) energy; moreover kinetic energy can be generated from potential energy. Unfortunately, unlike momentum, the kinetic energy of a collection of particles does not equal that of the center of mass - this is because kinetic energy depends quadratically rather than linearly on the velocity. The total kinetic energy does of course equal the sum of the individual particles' kinetic energies. Moreover, here too the decomposition (4.8) is useful:

$$\begin{aligned} K &= \sum_\alpha \frac{1}{2} m_\alpha (\boldsymbol{v}_{\mathrm{cm}} + \boldsymbol{v}_{\alpha,\mathrm{rel}}) \cdot (\boldsymbol{v}_{\mathrm{cm}} + \boldsymbol{v}_{\alpha,\mathrm{rel}}) & (4.11) \\ &= \sum_\alpha \frac{1}{2} m_\alpha v_{\mathrm{cm}}^2 + \sum_\alpha m_\alpha \boldsymbol{v}_{\mathrm{cm}} \cdot \boldsymbol{v}_{\alpha,\mathrm{rel}} + \sum_\alpha \frac{1}{2} m_\alpha v_{\alpha,\mathrm{rel}}^2 & (4.12) \\ &= K_{\mathrm{cm}} + K_{\mathrm{int}} & (4.13) \end{aligned}$$

Because the center of mass velocity is the same for all particles, it can be taken out of the sum in equation (4.12). Therefore, the first term equals $\frac{1}{2} M v_{\mathrm{cm}}^2 = K_{\mathrm{cm}}$, and in the second term we end up with the sum over *all* velocities relative to the center of mass - which is zero. We find that the total kinetic energy of a collection of particles equals the kinetic energy of the center of mass *plus* the total internal kinetic energy - which can change in both collisions and when potential energy gets converted into kinetic energy (or vice versa).

4.4 ROCKET SCIENCE*

Although designing a rocket that will follow a desired trajectory (say to Ceres, Pluto, or Planet Nine) with great accuracy is an enormous engineering challenge, the basic principle behind rocket propulsion is remarkably simple. It essentially boils down to conservation of momentum, or, equivalently, the observation that the velocity of center of mass of a system does not change if no external forces are acting on the system. To understand how a rocket works, imagine[1] the following experiment: you sit on a initially stationary cart with a large amount of small balls. You then pick up the balls one by one, and throw them all in the same direction with the same (preferably high) speed (relative to yourself and thus the cart). What will happen is that you, the cart, and the remaining balls slowly pick up speed, in the opposite direction from the one you're throwing the balls in. This is exactly what a rocket engine does: it thrusts out small particles (molecules, actually) at high velocities, gaining a small velocity itself in the opposite direction. Note that this is completely different from most other engines, which drive the rotation of wheels (that depend on friction to work) or propellers (that depend on drag to work).

4.4.1 ROCKET EQUATION

To understand what happens in our thought experiment, let's first consider the first ball you throw. Let's call the mass of yourself plus the cart M, the total mass of the balls m, and the (small) mass of a single ball $\mathrm{d}m$. If you trow the ball with a speed u (with respect to yourself), we can calculate your resulting speed in two ways:

1. The center of mass must remain stationary. Let's put $x_{\mathrm{cm}} = 0$. Before the throw, we then have $x_{\mathrm{ball}} \mathrm{d}m + x_{\mathrm{car}}(M+m) = 0$, whereas after the throw we have $-ut\,\mathrm{d}m + v_{\mathrm{car}} t(M+m) = 0$, or $v_{\mathrm{car}} = -u\,\mathrm{d}m/(M+m)$.
2. The total momentum must be conserved. Before the throw, the total momentum is zero, as nothing is moving. After the throw, we get: $p_{\mathrm{ball}} + p_{\mathrm{car}} = -u\,\mathrm{d}m + v_{\mathrm{car}}(M+m)$. Equating this to zero again gives $v_{\mathrm{car}} = -u\,\mathrm{d}m/(M+m)$.

Now for the second, third, etc. ball, the situation gets more complicated, as the car (including the ball that is about to be thrown) is already moving. Naturally, the center of mass of the car plus all the balls remains fixed, as does the total momentum of the car plus all the balls. However, to calculate how much extra speed the car picks up from the nth ball, it is easier to not consider the balls already thrown. Instead, we consider a car (including the remaining balls) that is already moving at speed v, and thus has total momentum $(M+m)v$. Throwing the next ball will reduce the mass of the car plus balls by $\mathrm{d}m$, and increase its velocity by $\mathrm{d}v$. Conservation of momentum then gives:

$$(M+m)v = (M+m-\mathrm{d}m)(v+\mathrm{d}v) + (v-u)\,\mathrm{d}m = (M+m)v + (M+m)\,\mathrm{d}v - u\,\mathrm{d}m, \tag{4.14}$$

[1] Or carry out, as you please.

Konstantin Eduardovich Tsiolkovsky (1857-1935) was a Russian rocket scientist, who is considered to be one of the pioneers of cosmonautics. Self-taught, Tsiolkovsky became interested in spaceflight both through 'cosmic' philosopher Nikolai Fyodorov and science-fiction author Jules Verne and considered the construction of a space elevator inspired by the then newly built Eiffel tower in Paris. Working as a teacher, he spent much of his free time on research, developing the rocket equation named after him (equation 4.16) as well as developing designs for rockets, including multi-stage ones. Tsiolkovsky also worked on designing airplanes and airships (dirigibles), but did not get support from the authorities to develop these further. He kept working on rockets though, while also continuing as a mathematics teacher. Only late in life did he receive recognition for his work at home (then the Soviet Union), but his ideas would go on to influence other rocket pioneers in both the Soviet and American space programs.

Figure 4.2: Konstantin Tsiolkovsky [14].

where we dropped the second-order term $\mathrm{d}m\,\mathrm{d}v$. Equation (4.14) can be rewritten to

$$(M+m)\,\mathrm{d}v = u\,\mathrm{d}m. \tag{4.15}$$

Note that here both u (the speed of each thrown ball) and M (the mass of yourself plus the car, or the shell of a rocket) are constants, whereas m changes, ending up at zero when you've thrown all your balls. To find the velocity of our car, we can integrate equation (4.15), but there is an important, and rather subtle, point to consider. The left-hand side of equation (4.15) applies to the car, but the right-hand side to the thrown ball, with a (positive) mass $\mathrm{d}m$. The mass m of the balls remaining in the car, however, has *decreased* by $\mathrm{d}m$, so if we wish to know the final velocity of the car, we need to include a minus sign on the right-hand side of equation (4.15). Dividing through by $M+m$ and integrating, we then obtain:

$$\Delta v = v_\mathrm{f} - v_0 = u\log\left(\frac{M+m_0}{M}\right), \tag{4.16}$$

where v_f is the final velocity of the car, and m_0 the initial total mass of all the balls. Equation (4.16) is known as the *Tsiolkovsky rocket equation*[2].

4.4.2 MULTI-STAGE ROCKETS

Because of the logarithmic factor in the Tsiolkovsky rocket equation, rockets need a lot of fuel compared to the mass of the object they intend to deliver (the payload - say a probe, or a capsule with astronauts). Even so, the effectiveness of rockets is limited. A fuel to payload ratio of 9:1 (already quite high) and an initial speed of zero gives a final speed $v_\mathrm{f} = u\log(10) \simeq 2.3u$, and increasing the ratio to 99:1 only doubles this result: $v_\mathrm{f} = u\log(100) \simeq 4.6u$. To get around these limitations and give rockets (or rather their payloads) the speed necessary to leave Earth, or even the solar system, rockets are built with multiple stages - essentially a number of rockets stacked one upon the next. If these stages all have the same fuel to payload ratio and exhaust velocity, the final velocity of the payload simply is that of a single stage times the number of stages n: $v_\mathrm{f} = nu\log(1+m_0/M)$. To see this, consider that the remaining stages are the payload of the current stage. Having multiple stages thus allows rockets to pick up speed more efficiently, essentially by shedding a part of the 'payload' (casing of an empty stage). For example, the Saturn V rocket that was used to send the Apollo astronauts to the moon had three stages, plus a small rocket engine on the capsule itself (used to break moon orbit and send the astronauts back to Earth), see figure 4.3.

4.4.3 IMPULSE

When you're crashing into something, there are two factors that determine how much your momentum changes: the amount of force acting on you, and the time the force is acting. The product is known as the

[2] Though Tsiolkovsky certainly deserves credit for his pioneering work, and he likely derived the equation independently, he was not the first to do so. Both the British mathematician William Moore in 1813 and the Scottish minister and mathematician William Leitch in 1861 preceded him.

Figure 4.3: Rockets and related spacecraft that took people to the moon in the late 1960's and early 1970's [15]. (a) Aerial view of a Saturn V rocket on its launchpad. This rocket carries the Apollo 15, the fourth mission to make it to the moon. The three rocket stages are separated by rings around the engine of the next stage. The total height of the rocket at launch was 110.6 m; it had a total mass of 2.97 million kg, and could take a payload of 140000 kg to low-Earth orbit or a payload of 48600 kg to the moon. (b) View from the launch tower of the Saturn V carrying Apollo 11 (the famous first mission to the moon in 1969) at ignition. The little rocket on top was to be used for an emergency escape of the manned module immediately below if anything went wrong at launch. The manned 'command module' is the little conical structure; the cylindrical structure directly below it contained its engine, and the conical part below that contained the lunar lander (figure d). (c) Jettisoned third stage of the Saturn V rocket that carried the Apollo 17 mission (the sixth and last(!) to make it to the moon in 1972). The empty space at the front contained the lunar lander module at launch. (d) Lunar lander of the 1969 Apollo 11 mission, photographed from the command module after separation. This module contained two rockets: one to slow the descent to the moon on the lower part, and one to return to moon orbit with just the upper part. The lower part of the lander remains on the moon and was photographed there in 2012 by the Lunar Reconnaisance Orbiter, an unmanned spacecraft in Moon orbit.

impulse, which by Newton's second law equals the change in momentum:

$$J = \Delta p = \int F(t)\,dt. \tag{4.17}$$

The *specific impulse*, defined as $I_{sp} = J/m_{propellant}$, or the impulse per unit mass of fuel, is a measure of the efficiency of jet engines and rockets.

4.5 COLLISIONS

Collisions occur when two (or more) particles hit each other. During a collision, those particles exert forces on each other, but in general, there are no external forces acting on the system consisting of the colliding particles. Consequently, the total momentum of all particles involved in the collision is conserved. Typically, we know the initial velocities and the masses of the particles, and want to calculate their final velocities - though of course you could also do it the other way around.

Although the total momentum of colliding particles is conserved, the total (kinetic) energy of all particles typically is not, as irreversible processes (such as plastic deformations of the particles) occur that are associated with nonconservative forces. In the special case that kinetic energy is conserved in a collision, we call the collision (totally) *elastic*. All other collisions are called *inelastic*, with the extreme case of a *totally inelastic* collision, in which the colliding objects stick together.

4.6 TOTALLY INELASTIC COLLISIONS

For the case of two particles colliding totally inelastically, conservation of momentum gives:

$$m_1 \boldsymbol{v}_1 + m_2 \boldsymbol{v}_2 = (m_1 + m_2)\boldsymbol{v}_f. \tag{4.18}$$

If the masses and initial velocities of the particles are known, calculating the final velocity of the composite particle is thus straightforward.

4.6.1 WORKED EXAMPLE: BIKE CRASH

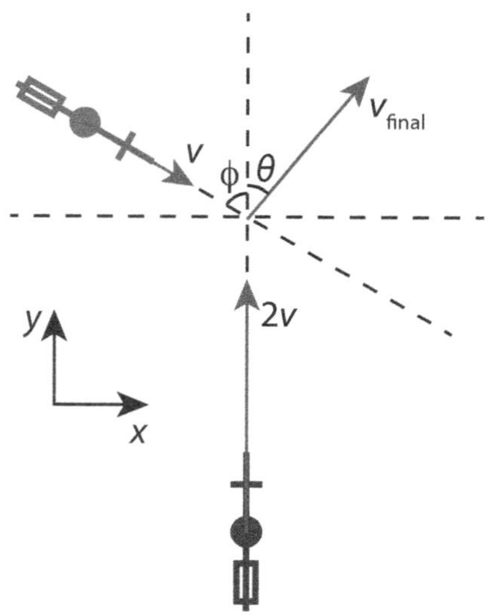

Figure 4.4: Two colliding cyclists.

You're late for class and it's raining to boot, so you cycle as fast as you possibly can, without paying attention in which direction you're going. A classmate, similarly late, comes towards you from a side street that makes an angle ϕ with yours. When your streets cross, you crash into each other, moving together in a big clutch of people and bikes, see figure 4.4. Suppose you're about equally heavy, but you're the faster biker, with initially twice the speed of your classmate. Find the velocity (i.e., magnitude and direction) you and your classmate have immediately after the collision.

SOLUTION

Let's call your (and your classmate's) mass m, your initial speed $2v$ (so your classmate's speed is v), and your combined final speed v_f, with an angle θ with your initial direction. After the collision you move as one object, so the collision is completely inelastic. During the collision we have conservation of momentum in both the x and y directions, which gives:

$$0 + v\sin\phi = v_f \sin\theta, \tag{4.19}$$

$$2v - v\cos\phi = v_f \cos\theta. \tag{4.20}$$

We need to solve for both v_f and θ. To eliminate θ, we square both (4.19) and (4.20) and add them, which gives

$$v_f^2 = v^2 \sin^2\phi + v^2(2-\cos\phi)^2 = v^2(5-4\cos\phi), \tag{4.21}$$

or $v_f = v\sqrt{5-4\cos\phi}$. To get θ, we divide (4.19) by (4.20), which gives

$$\tan\theta = \frac{\sin\phi}{2-\cos\phi}. \tag{4.22}$$

We can easily check that these answers make sense when $\phi = 0$, which gives $v_f = v$ and $\theta = 0$, as it should.

4.7 TOTALLY ELASTIC COLLISIONS

For a totally elastic collision, we can invoke both conservation of momentum and (by definition of a totally elastic collision) of kinetic energy. We also have an additional variable, as compared to the totally inelastic case, because in this case the objects do not stick together and thus get different end speeds. The two equations governing a totally elastic collision are:

$$m_1 \boldsymbol{v}_{1,i} + m_2 \boldsymbol{v}_{2,i} = m_1 \boldsymbol{v}_{1,f} + m_2 \boldsymbol{v}_{2,f}, \tag{4.23}$$

for momentum conservation, and

$$\frac{1}{2}m_1 v_{1,i}^2 + \frac{1}{2}m_2 v_{2,i}^2 = \frac{1}{2}m_1 v_{1,f}^2 + \frac{1}{2}m_2 v_{2,f}^2 \tag{4.24}$$

for kinetic energy conservation.

When the collision occurs in one dimension, we can combine equations (4.23) and (4.24) to calculate the final velocities as functions of the initial ones. We first rewrite the two equations so that everything associated with particle 1 is on the left, and the terms for particle 2 are on the right:

$$m_1(v_{1,i} - v_{1,f}) = m_2(v_{2,f} - v_{2,i}), \tag{4.25}$$

and

$$m_1(v_{1,i}^2 - v_{1,f}^2) = m_2(v_{2,f}^2 - v_{2,i}^2). \tag{4.26}$$

We can expand the terms in parentheses in equation (4.26), which gives:

$$m_1(v_{1,i} - v_{1,f})(v_{1,i} + v_{1,f}) = m_2(v_{2,f} - v_{2,i})(v_{2,f} + v_{2,i}). \tag{4.27}$$

Dividing equation (4.27) by equation (4.25), we get a relation between the velocities alone:

$$v_{1,i} + v_{1,f} = v_{2,i} + v_{2,f}. \tag{4.28}$$

From equation (4.28) we can isolate $v_{2,f}$ and substitute back in (4.25) to find $v_{1,f}$ in terms of the initial velocities:

$$v_{1,f} = \frac{m_1 - m_2}{m_1 + m_2} v_{1,i} + 2\frac{m_2}{m_1 + m_2} v_{2,i}. \tag{4.29}$$

Naturally, we could just as well have calculated $v_{2,f}$, the equation for which is just (4.29) with the 1's and 2's swapped:

$$v_{2,f} = 2\frac{m_1}{m_1 + m_2} v_{1,i} + \frac{m_2 - m_1}{m_1 + m_2} v_{2,i}. \tag{4.30}$$

We note that in the limit case that $m_2 >> m_1$, v_2 is hardly affected, and $v_{1,f} \simeq -v_{1,i} + 2v_{2,f}$.

4.8 ELASTIC COLLISIONS IN THE COM FRAME

Equations (4.29) and (4.30) give the final velocities of two particles after a totally elastic collision. We did the calculation in the lab frame, i.e., from the point of view of a stationary observer. We could of course just as well have done the calculation in the center-of-mass (COM) frame of section 4.3. Within that frame, as we'll see below, the relation between the initial and final velocities in an elastic collision is much simpler than in the lab frame.

We will use equation (4.8) to calculate velocities in the COM frame. For notational simplicity, we'll work in one dimension, and use an overbar to indicate velocities in the COM frame, so we get

$$v_i = v_{\text{cm}} + \bar{v}_i$$

for each particle i. The velocity of the center of mass is simply the time derivative of its position. For two particles, it is given by

$$v_{\text{cm}} = \frac{m_1 v_1 + m_2 v_2}{m_1 + m_2}. \tag{4.31}$$

The velocities of the two particles in the COM frame is then

$$\bar{v}_1 = v_1 - v_{\text{cm}} = \frac{m_2}{m_1 + m_2}(v_1 - v_2) = \frac{m_2}{m_1 + m_2} v_{\text{rel}}, \tag{4.32}$$

$$\bar{v}_2 = v_2 - v_{\text{cm}} = \frac{m_1}{m_1 + m_2}(v_2 - v_1) = -\frac{m_1}{m_1 + m_2} v_{\text{rel}}, \tag{4.33}$$

where $v_{\text{rel}} = v_1 - v_2 = \bar{v}_1 - \bar{v}_2$ is the *relative velocity* of the two particles[3]. Note that it does not matter whether we calculate the relative velocity in the lab or COM frame. Equations (4.32) and (4.33) have a nice symmetry in their velocity components, but not in their mass components. The symmetry is more complete if instead of velocities, we consider momenta in the COM frame, where $\bar{p} = m\bar{v}$:

$$\bar{p}_1 = m_1 \bar{v}_1 = \frac{m_1 m_2}{m_1 + m_2} v_{\text{rel}} = \mu v_{\text{rel}}, \tag{4.34}$$

$$\bar{p}_2 = m_2 \bar{v}_2 = -\frac{m_1 m_2}{m_1 + m_2} v_{\text{rel}} = -\mu v_{\text{rel}}, \tag{4.35}$$

where we introduced a new variable μ, the *reduced mass*

$$\mu = \frac{m_1 m_2}{m_1 + m_2}. \tag{4.36}$$

Clearly the total momentum in the center of mass frame is zero[4] (as it should be), both before and after a collision, and is thus conserved. To find out what happens with the relative velocity in an elastic collision, we invoke conservation of kinetic energy, which we calculate using $K = \frac{1}{2}mv^2 = p^2/2m$:

$$\frac{\bar{p}_{1,\text{i}}^2}{2m_1} + \frac{\bar{p}_{2,\text{i}}^2}{2m_2} = \frac{\bar{p}_{1,\text{f}}^2}{2m_1} + \frac{\bar{p}_{2,\text{f}}^2}{2m_2} \tag{4.37}$$

$$\frac{\mu^2 \bar{v}_{\text{rel,i}}^2}{2m_1} + \frac{\mu^2 \bar{v}_{\text{rel,i}}^2}{2m_2} = \frac{\mu^2 \bar{v}_{\text{rel,f}}^2}{2m_1} + \frac{\mu^2 \bar{v}_{\text{rel,f}}^2}{2m_2} \tag{4.38}$$

$$\bar{v}_{\text{rel,i}}^2 = \bar{v}_{\text{rel,f}}^2. \tag{4.39}$$

We find that either $\bar{v}_{\text{rel,f}} = \bar{v}_{\text{rel,i}}$, in which case there would be no collision (as nothing changes), or $\bar{v}_{\text{rel,f}} = -\bar{v}_{\text{rel,i}}$, which means that in an elastic collision in the COM frame, the velocities (and momenta) of the colliding particles reverse. We get:

$$\bar{v}_{1,\text{f}} = -\bar{v}_{1,\text{i}} = -\frac{m_2}{m_1 + m_2}(v_{1,\text{i}} - v_{2,\text{i}}) = -\frac{m_2}{m_1 + m_2} v_{\text{rel,i}}, \tag{4.40}$$

$$\bar{v}_{2,\text{f}} = -\bar{v}_{2,\text{i}} = -\frac{m_1}{m_1 + m_2}(v_{2,\text{i}} - v_{1,\text{i}}) = \frac{m_1}{m_1 + m_2} v_{\text{rel,i}}. \tag{4.41}$$

We can of course transform these expressions back to the lab frame by adding the center of mass velocity (4.31), which gives equations (4.29) and (4.30), the same as our calculation in the lab frame.

[3] Equations (4.32) and (4.33) hold in multiple dimensions as well.

[4] Since the total momentum in the COM frame is zero, the frame is sometimes also referred to as the zero-momentum frame.

4.9 PROBLEMS

4.1 **Celestial centers of mass** We say that the Moon orbits the Earth, because the Earth's gravity pulls on the Moon, causing it to orbit. However, by Newton's third law, the Moon exerts a force back on the Earth. Therefore, the Earth should move in response to the Moon. Thus a more accurate statement would be that the Moon and the Earth both orbit the center of mass of the Earth-Moon system. Useful values: $M_E = 5.97 \cdot 10^{24}$ kg, $R_E = 6.37 \cdot 10^6$ m, $R_{E,orbit} = 1.50 \cdot 10^{11}$ m, $M_M = 7.35 \cdot 10^{22}$ kg, $R_M = 1.74 \cdot 10^6$ m, $R_{M,orbit} = 3.84 \cdot 10^8$ m, $M_S = 1.99 \cdot 10^{30}$ kg, $R_S = 6.96 \cdot 10^8$ m.

(a) Find the center of mass of the Earth-Moon system (in an appropriately chosen, and clearly defined, coordinate system). Does this center of mass lie inside the Earth?

(b) Find the location of the center of mass of the Earth-Moon-Sun system during a full Moon.

(c) Find the location of the center of mass of the Earth-Moon-Sun system when the Moon is in its first quarter.

4.2 A shell is shot with an initial velocity of 20 m/s, at an angle of 60° with the horizontal. At the top of the trajectory, the shell explodes into two fragments of equal mass. One fragment, whose speed immediately after the explosion is zero, drops to the ground vertically. How far from the gun does the other fragment land (assuming no air drag and level terrain)?

4.3 Two cannonballs with masses m_1 and m_2 are simultaneously fired from two cannons situated a distance L apart.

(a) Find the equations of motion for the horizontal and vertical components of the vector describing the center of mass of the cannonballs.

(b) Show that the motion of the center of mass is a parabola through space.

4.4 **Center of mass of some solid objects**

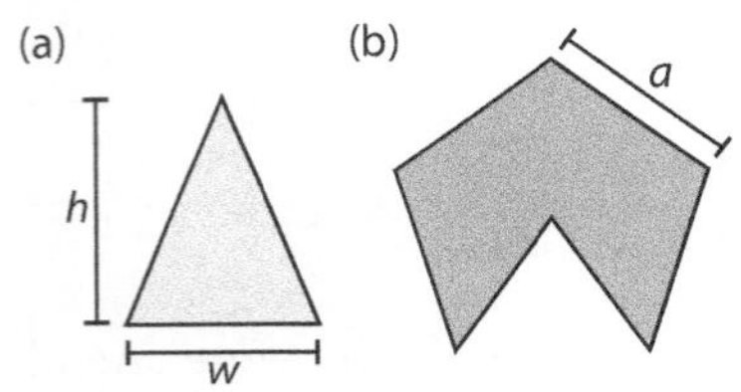

(a) Find the center of mass of an isosceles triangle with a base width w and height h (see figure a).

(b) Find the center of mass of a pentagon with five equal sides of length a, but with one triangle missing (see figure b). Hint: use your result from (a).

(c) Find the position of the center of mass of a semicylinder (half of a solid cylinder, i.e., a solid cylinder sliced in two along a plane containing its symmetry axis). *Hint*: first calculate the center of mass of half a solid disk.

4.5 A dog (black dot in the sketch below) of mass m stands at the end of a boat of mass M and length L at an initial distance D from the shore. The dog then walks to the other end of the boat and stops there. Assuming no friction between the boat and the water, how far is the dog then from the shore? (*Hint*: what is conserved?).

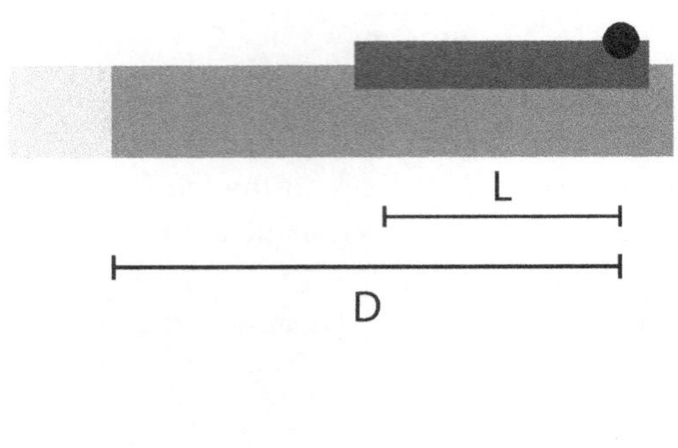

Figure 4.5: Serena Williams serving at the 2008 Wimbledon championships [18].

4.6 [For optional section 4.4] Every point in a tennis match stars with one of the players serving. The most commonly used service involves the player tossing the ball in the air and hitting it with their racket. To get the ball to move as fast as possible, players commonly swing the racket to give it a large momentum, and deliver a maximal impulse to the ball.

Figure 4.5 shows Serena Williams serving during the 2008 Wimbledon championships. Williams is widely regarded as one of the best women tennis players and holds the record of most aces (scoring a point from a serve without the opponent touching the ball) by a female player at a Grand Slam tournament.

Williams is 175 cm tall. As you can see in the figure, at the top of its trajectory, the ball is about twice Williams' height above the ground. Also, as the span of people's arms is about the same as their height, and the shoulders of an adult are at about 5/6th of their height, we can estimate Williams' arms to be about 75 cm long and her shoulders to be at 145 cm above the ground. The distance between the point where a player holds the racket and where they hit the ball is typically about 40 cm.

(a) Find the speed of the ball coming down at the moment Williams hits it, assuming that she hits it with a fully upward stretched arm.

(b) Williams's personal record for serve speed (speed of the ball after it was hit by the racket) is 207 km/h. Determine the impulse she delivered with her racket on the 58.0 g tennis ball as she hit it.

(c) Assuming a typical racket weight of 360 g, calculate the change in speed of the racket from just before to just after Williams hit the ball.

(d) Calculate the magnitude of the force on Williams' hand at the moment she hits the ball with her racket.

4.7 A 2.75 g bullet embeds itself in a 1.50 kg block, which is attached to a spring of force constant 850 N/m. The maximum compression of the spring is 4.30 cm.

(a) Determine the initial speed of the bullet.

(b) Find the time it takes the bullet-block system to come to rest.

4.8 **Head-on collision between two balls** A ball of mass m has velocity v when it makes a head-on collision with another ball of mass M that is originally at rest. After the collision the ball of mass m rebounds straight back along its path with $2/3^{\text{rd}}$ of its initial kinetic energy. We assume that the collision is totally elastic.

(a) Sketch the situation before and after the collision, indicating directions of velocity, and values (if known, give symbols otherwise).

(b) Write down all applicable conservation laws for this case.

(c) From the conservation laws, solve for the mass M of the ball that is initially at rest.

(d) Also solve for the velocity of that ball after the collision.

4.9 A small ball of mass m is aligned above a larger ball of mass M with a slight separation, and the two are dropped simultaneously from a height h. Assume the radii of the two balls and the initial separation are negligible compared to h.

(a) If the larger ball rebounds elastically from the floor and then the small ball rebounds elastically from the larger ball, what value of m (as a fraction of M) results in the larger ball stopping when it collides with the small ball?

(b) What height does the small ball then reach?

5

ROTATIONAL MOTION, TORQUE AND ANGULAR MOMENTUM

5.1 ROTATION BASICS

So far, we've been looking at motion that is easily described in Cartesian coordinates, often moving along straight lines. Such kind of motion happens a lot, but there is a second common class as well: rotational motion. It won't come as a surprise that to describe rotational motion, polar coordinates (or their 3D counterparts the cylindrical and spherical coordinates) are much handier than Cartesian ones[1]. For example, if we consider the case of a disk rotating with a uniform velocity around its center, the easiest way to do so is to specify over how many degrees (or radians) a point on the boundary advances per second. Compare this to linear motion - that is specified by how many meters you advance in the linear direction per second, which is the speed (with dimension L/T). The change of the angle per second gives you the angular speed ω, where a counterclockwise rotation is taken to be in the positive direction. The angular speed has dimension $1/T$, so it is a *frequency*. It is measured in degrees per second or radians per second. If the angle at a point in time is denoted by $\theta(t)$, then obviously $\omega = \dot{\theta}$, just like $v = \dot{x}$ in linear motion.

In three dimensions, $\boldsymbol{\omega}$ becomes a vector, where the magnitude is still the rotational speed, and the direction gives you the direction of the rotation, by means of a right-hand rule: rotation is in the plane perpendicular to $\boldsymbol{\omega}$, and in the direction the fingers of your right hand point if your thumb points along $\boldsymbol{\omega}$ (this gives $\boldsymbol{\omega}$ in the positive $\hat{\boldsymbol{z}}$ direction for rotational motion in the xy plane).

Going back to 2D for the moment, let's call the angular position $\theta(t)$, then

$$\omega = \frac{\mathrm{d}\theta}{\mathrm{d}t} = \dot{\theta}. \tag{5.1}$$

If we want to know the position $\boldsymbol{r}$ in Cartesian coordinates, we can simply use the normal conversion from polar to Cartesian coordinates, and write

$$\boldsymbol{r}(t) = r\cos(\omega t)\hat{\boldsymbol{x}} + r\sin(\omega t)\hat{\boldsymbol{y}} = r\hat{\boldsymbol{r}}, \tag{5.2}$$

where r is the distance to the origin. Note that $\boldsymbol{r}$ points in the direction of the polar unit vector $\hat{\boldsymbol{r}}$. Equation (5.2) gives us an interpretation of ω as a frequency: if we consider an object undergoing uniform rotation (i.e., constant radius and constant velocity), in its x and y-directions it oscillates with frequency ω.

As long as our motion remains purely rotational, the radial distance r does not change, and we can find the linear velocity by taking the time derivative of (5.2):

$$\boldsymbol{v}(t) = \dot{\boldsymbol{r}}(t) = -\omega r\sin(\omega t)\hat{\boldsymbol{x}} + \omega r\cos(\omega t)\hat{\boldsymbol{y}} = \omega r\hat{\boldsymbol{\theta}}, \tag{5.3}$$

so in particular we have $v = \omega r$. Note that both v and ω denote instantaneous speeds, and equation (5.2) only holds when ω is constant. However, the relation $v = \omega r$ always holds. To see that this is true, express θ in radians, $\theta = s/r$, where s is the distance traveled along the rotation direction. Then

$$\omega = \frac{\mathrm{d}\theta}{\mathrm{d}t} = \frac{1}{r}\frac{\mathrm{d}s}{\mathrm{d}t} = \frac{v}{r}. \tag{5.4}$$

[1] If you need a refresher on polar coordinates, or are unfamiliar with polar basis vectors, check out appendix A.2.

In three dimensions, we find

$$\boldsymbol{v} = \boldsymbol{\omega} \times \boldsymbol{r}, \tag{5.5}$$

where $\boldsymbol{r}$ points from the rotation axis to the rotating point.

Unlike in linear motion, in rotational motion there is always acceleration, even if the rotational velocity ω is constant. This acceleration originates in the fact that the *direction* of the (linear) velocity always changes as points revolve around the center, even if its magnitude, the net linear speed, is constant. In that special case, taking another derivative gives us the linear acceleration, which points towards the center of rotation:

$$\boldsymbol{a}(t) = \ddot{\boldsymbol{r}}(t) = -\omega^2 r \cos(\omega t)\hat{\boldsymbol{x}} + \omega^2 r \sin(\omega t)\hat{\boldsymbol{y}} = -\omega^2 r \hat{\boldsymbol{r}}. \tag{5.6}$$

In section 5.2 below we will use equation (5.6) in combination with Newton's second law of motion to calculate the net centripetal force required to maintain rotation at a constant rate.

Of course the angular velocity ω need not be constant at all. If it is not, we can define an *angular acceleration* by taking its time derivative:

$$\alpha = \frac{\mathrm{d}\omega}{\mathrm{d}t} = \ddot{\theta}, \tag{5.7}$$

or in three dimensions, where $\boldsymbol{\omega}$ is a vector:

$$\boldsymbol{\alpha} = \frac{\mathrm{d}\boldsymbol{\omega}}{\mathrm{d}t}. \tag{5.8}$$

Note that when $\boldsymbol{\alpha}$ is parallel to $\boldsymbol{\omega}$, it simply represents a change in the rotation rate (i.e., a speeding up/slowing down of the rotation), but when it is not, it also represents a change of the plane of rotation.

In both two and three dimensions, a change in rotation rate causes the linear acceleration to have a component in the tangential direction in addition to the radial acceleration (5.6). The tangential component of the acceleration is given by the derivative of the linear velocity:

$$\boldsymbol{a}_\mathrm{t} = \frac{\mathrm{d}\boldsymbol{v}}{\mathrm{d}t} = r\frac{\mathrm{d}\boldsymbol{\omega}}{\mathrm{d}t} = r\boldsymbol{\alpha}. \tag{5.9}$$

In two dimensions, $\boldsymbol{a}_\mathrm{t}$ points along the $\pm\hat{\boldsymbol{\theta}}$ direction.

Naturally, there are even more complicated possibilities - the radius of the rotational motion can change as well. We'll look at that case in more detail in chapter 6, but first we consider 'pure' rotations, where the distance to the rotation axis is fixed.

5.2 CENTRIPETAL FORCE

When you're rotating at constant angular velocity, the magnitude of your velocity is always the same, but its direction constantly changes - so you're constantly undergoing an acceleration, as indicated in equation (5.6). Therefore there must be a net force acting on you. We can calculate that net force using Newton's second law of motion. It is known as the *centripetal force* and given by:

$$\boldsymbol{F}_\mathrm{cp} = m\boldsymbol{a} = -\frac{mv^2}{r}\hat{\boldsymbol{r}} = -m\omega^2 r\hat{\boldsymbol{r}}. \tag{5.10}$$

'Centripetal' means 'center-seeking' (from Latin 'centrum' = center and 'petere' = to seek). It is important to remember that this is a net resulting force, not a 'new' force like that exerted by gravity or a compressed spring. Equation (5.10) is after all just a special case of Newton's second law of motion.

5.3 TORQUE

Anyone who has ever used a lever - that is everyone, presumably - knows how useful they are at augmenting force: you push with a small force at the long end, to produce a large force at the short end, and make the crank turn, stone lift or bottle cap pop off. If the force is at straight angles with the lever arm (the line connecting the point at which you exert the force to the pivot around which your lever rotates), the effect is largest. In that case we define the *torque* (or *moment of force*) τ as the product of the force and the lever arm length. As only the perpendicular component of the force counts (you'll simply push or pull on your lever with a parallel component, not make it turn), in a vector setting you need to project on that perpendicular component, so if $\boldsymbol{r}$ (from the pivot to the point at which the force acts) makes an angle θ with the force vector $\boldsymbol{F}$, the magnitude of the torque becomes $\tau = rF\sin\theta$. That's exactly the magnitude of the cross product of $\boldsymbol{r}$ and $\boldsymbol{F}$, which

Object	Rotation axis	Moment of inertia
Stick	Center, perpendicular to stick	$\frac{1}{12}ML^2$
Stick	End, perpendicular to stick	$\frac{1}{3}ML^2$
Cylinder, hollow	Center, parallel to axis	MR^2
Cylinder, solid	Center, parallel to axis	$\frac{1}{2}MR^2$
Sphere, hollow	Any axis through center	$\frac{2}{3}MR^2$
Sphere, solid	Any axis through center	$\frac{2}{5}MR^2$
Planar object, size $a \times b$	Axis through center, in plane, parallel to side with length a	$\frac{1}{12}Mb^2$
Planar object, size $a \times b$	Axis through center, perpendicular to plane	$\frac{1}{12}M(a^2+b^2)$

Table 5.1: Moments of inertia for some common objects, all with total mass M and length L / radius R.

also has directional information - useful as a torque can be clockwise or counterclockwise. In general we'll therefore define the torque by:

$$\boldsymbol{\tau} = \boldsymbol{r} \times \boldsymbol{F}, \tag{5.11}$$

which makes a counterclockwise torque positive, in correspondence with the definition of the rotation vector $\boldsymbol{\omega}$.

5.4 Moment of inertia

Suppose we have a mass m at the end of a massless stick of length r, rotating around the other end of the stick. If we want to increase the rotation rate, we need to apply a tangential acceleration $\boldsymbol{a}_\mathrm{t} = r\boldsymbol{\alpha}$, for which by Newton's second law of motion we need a force $\boldsymbol{F} = m\boldsymbol{a}_\mathrm{t} = mr\boldsymbol{\alpha}$. This force in turn generates a torque of magnitude $\tau = r \cdot F = mr^2\alpha$. The last equality is reminiscent of Newton's second law of motion, but with force replaced by torque, acceleration by angular acceleration, and mass by the quantity mr^2. In analog with mass representing the inertia of a body undergoing linear acceleration, we'll identify this quantity as the inertia of a body undergoing rotational acceleration, which we'll call the *moment of inertia* and denote by I:

$$\boldsymbol{\tau} = I\boldsymbol{\alpha}. \tag{5.12}$$

Equation (5.12) is the rotational analog of Newton's second law of motion. By extending our previous example, we can find the moment of inertia of an arbitrary collection of particles of masses m_α and distances to the rotation axis r_α (where α runs over all particles), and write:

$$I = \sum_\alpha m_\alpha r_\alpha^2, \tag{5.13}$$

which like the center of mass in section 4.1 easily generalizes to continuous objects as[2]

$$I = \int_V (\boldsymbol{r} \cdot \boldsymbol{r})\rho \,\mathrm{d}V = \int_V \rho r^2 \,\mathrm{d}V. \tag{5.14}$$

Note that it matters where we choose the rotation axis. For example, the moment of inertia of a rod of length L and mass m around an axis through its center perpendicular to the rod is $\frac{1}{12}mL^2$, whereas the moment of inertia around an axis perpendicular to the rod but located at one of its ends is $\frac{1}{3}mL^2$. Also, moments of inertia are different for hollow and solid objects - a hollow sphere of mass m and radius R has $\frac{2}{3}mR^2$ whereas a solid sphere has $\frac{2}{5}mR^2$, and for hollow and solid cylinders (or hoops and disks) around the long axis through the center we find mr^2 and $\frac{1}{4}mr^2$ respectively. These and some other examples are listed in table 5.1. Below we'll relate the moment of inertia to the kinetic energy of a moving-and-rolling object, but first we present two handy theorems that will help in calculating them.

Theorem 5.1 (Parallel axis theorem). *If the moment of inertia of a rigid body about an axis through its center of mass is given by I_cm, then the moment of inertia around an axis parallel to the original axis and separated from it by a distance d is given by*

$$I = I_\mathrm{cm} + md^2, \tag{5.15}$$

[2] Like the one- and two-dimensional analogs of the center of mass of a continuous object (4.3), there are one- and two-dimensional analogs of (5.14), which you get by replacing ρ with λ or σ and $\mathrm{d}V$ by $\mathrm{d}x$ or $\mathrm{d}A$, respectively.

where m is the object's mass.

Proof. Choose coordinates such that the center of mass is at the origin, and the original axis coincides with the $\hat{\mathbf{z}}$ axis. Denote the position of the point in the xy plane through which the new axis passes by $\mathbf{d}$, and the distance from that point for any other point in space by $\mathbf{r}_d$, such that $\mathbf{r} = \mathbf{d} + \mathbf{r}_d$. Now calculate the moment of inertia about the new axis through $\mathbf{d}$:

$$\begin{aligned} I &= \int_V (\mathbf{r}_d \cdot \mathbf{r}_d)\rho \, \mathrm{d}V \\ &= \int_V (\mathbf{r}\cdot\mathbf{r} + \mathbf{d}\cdot\mathbf{d} - 2\mathbf{d}\cdot\mathbf{r})\rho \, \mathrm{d}V \\ &= I_{\mathrm{cm}} + md^2 - 2\mathbf{d}\cdot\int_V \mathbf{r}\rho \, \mathrm{d}V. \end{aligned} \tag{5.16}$$

Here $d^2 = \mathbf{d}\cdot\mathbf{d}$. The last integral in the last line of (5.16) is equal to the position of the center of mass, which we chose to be at the origin, so the last term vanishes, and we arrive at (5.15). □

Note that the first two lines of table 5.1 (moments of inertia of a stick) satisfy the perpendicular-axis theorem.

Theorem 5.2 (Perpendicular axis theorem). *If a rigid object lies entirely in a plane, and the moments of inertia around two perpendicular axes x and y in that plane are I_x and I_y, respectively, then the moment of inertia around the axis z perpendicular to the plane and passing through the intersection point, is given by*

$$I_z = I_x + I_y. \tag{5.17}$$

Proof. We simply calculate the moment of inertia around the z-axis (where A is the area of the object, and σ the mass per unit area):

$$I_z = \int_A (x^2 + y^2)\sigma \, \mathrm{d}A = \int_A x^2\sigma \, \mathrm{d}A + \int_A y^2\sigma \, \mathrm{d}A = I_y + I_x. \tag{5.18}$$

□

Note that the last two lines of table 5.1 (moments of inertia of a thin planar rectangle) satisfy the parallel axis theorem.

5.5 KINETIC ENERGY OF ROTATION

Naturally, a rotating object has kinetic energy - its parts are moving after all (even if they're just rotating around a fixed axis). The total kinetic energy of rotation is simply the sum of the kinetic energies of all rotating parts, just like the total translational kinetic energy was the sum of the individual kinetic energies of the constituent particles in section 4.5. Using that $v = \omega r$, we can write for a discrete collection of particles:

$$K_{\mathrm{rot}} = \sum_\alpha \frac{1}{2} m_\alpha v_\alpha^2 = \sum_\alpha \frac{1}{2} m_\alpha r_\alpha^2 \omega^2 = \frac{1}{2} I\omega^2, \tag{5.19}$$

by the definition (5.13) of the moment of inertia I. Analogously we find for a continuous object, using (5.14):

$$K_{\mathrm{rot}} = \int_V \frac{1}{2} v^2 \rho \, \mathrm{d}V = \int_V \frac{1}{2}\omega^2 r^2 \rho \, \mathrm{d}V = \frac{1}{2} I\omega^2, \tag{5.20}$$

so we arrive at the general rule:

$$K_{\mathrm{rot}} = \frac{1}{2} I\omega^2. \tag{5.21}$$

Naturally, the work-energy theorem (eq. 3.9) still holds, so we can use it to calculate the work necessary to effect a change in rotational velocity, which by equation (5.12) can also be expressed in terms of the torque (in 2D):

$$W = \Delta K_{\mathrm{rot}} = \frac{1}{2} I(\omega_{\mathrm{final}}^2 - \omega_{\mathrm{initial}}^2) = \int_{\theta_{\mathrm{initial}}}^{\theta_{\mathrm{final}}} \tau \, \mathrm{d}\theta. \tag{5.22}$$

5.6 ANGULAR MOMENTUM

In analogy with the definition of torque, $\boldsymbol{\tau} = \boldsymbol{r} \times \boldsymbol{F}$ as the rotational counterpart of the force, we define the *angular momentum* $\boldsymbol{L}$ as the rotational counterpart of momentum:

$$\boldsymbol{L} = \boldsymbol{r} \times \boldsymbol{p}. \tag{5.23}$$

For a rigid body rotating around an axis of symmetry, the angular momentum is given by

$$\boldsymbol{L} = I\boldsymbol{\omega}, \tag{5.24}$$

where I is the moment of inertia of the body with respect to the symmetry axis around which it rotates. Equation (5.24) also holds for a collection of particles rotating about a symmetry axis through their center of mass, as readily follows from (5.13) and (5.23). However, it does not hold in general, as in general, $\boldsymbol{L}$ does not have to be parallel to $\boldsymbol{\omega}$. For the general case, we need to consider a moment of inertia tensor $\boldsymbol{I}$ (represented as a 3×3 matrix) and write $\boldsymbol{L} = \boldsymbol{I} \cdot \boldsymbol{\omega}$. We'll consider this case in more detail in section 7.3.

5.7 CONSERVATION OF ANGULAR MOMENTUM

Given that the torque is the rotational analog of the force, and the angular momentum is that of the linear momentum, it will not come as a surprise that Newton's second law of motion has a rotational counterpart that relates the net torque to the time derivative of the angular momentum. To see that this is true, we simply calculate that time derivative:

$$\frac{\mathrm{d}\boldsymbol{L}}{\mathrm{d}t} = \frac{\mathrm{d}\boldsymbol{r}}{\mathrm{d}t} \times \boldsymbol{p} + \boldsymbol{r} \times \frac{\mathrm{d}\boldsymbol{p}}{\mathrm{d}t} = \boldsymbol{\tau}, \tag{5.25}$$

because $\dot{\boldsymbol{r}} \times \boldsymbol{p} = \boldsymbol{v} \times m\boldsymbol{v} = 0$. Some texts even use equation (5.25) as the definition of torque and work from there. Note that in the case that there is no external torque, we arrive at another conservation law:

Theorem 5.3 (Law of conservation of angular momentum). *When no external torques act on a rotating object, its angular momentum is conserved.*

Conservation of angular momentum is why a rolling hoop keeps rolling, and why a balancing a bicycle is relatively easy once you go fast enough.

What about collections of particles? Here things are a little more subtle. Writing $\boldsymbol{L} = \sum_i \boldsymbol{L}_i$ and again taking the derivative, we arrive at

$$\frac{\mathrm{d}\boldsymbol{L}}{\mathrm{d}t} = \sum_i \boldsymbol{r}_i \times \boldsymbol{F}_i = \sum_i \boldsymbol{\tau}_i. \tag{5.26}$$

Now the sum on the right hand side of (5.26) includes both external torques exerted on the system, and internal torques exerted by the particles on each other. When we discussed conservation of linear momentum, the internal momenta all canceled pairwise because of Newton's third law of motion. For torques this is not necessarily true, and we need the additional condition that the internal forces between two particles act along the line connecting those particles - then the internal torques are zero, and equation (5.25) holds for the collection as well. Consequently, if the net external torque is zero, angular momentum is again conserved.

5.8 ROLLING AND SLIPPING MOTION

When you slide an object over a surface (say, a book over a table), it will typically slow down quickly, due to frictional forces. When you do the same with a round object, like a water bottle, it may initially slide a little (especially if you push it hard), but will quickly start to rotate. You can easily check that when rotating, the object looses much less kinetic energy to work than when sliding - take the same water bottle, either on its bottom (sliding only) or on its side (a little sliding plus rolling), push it with the same initial force, and let go: the rolling bottle gets much further. However, somewhat ironically, the bottle can only roll thanks to friction. To start rolling, it needs to change its angular momentum, which requires a torque, which is provided by the frictional force acting on the bottle.

When a bottle (or ball, or any round object) rolls, the instantaneous speed of the point touching the surface over which it rolls is zero. Consequently, its rotational speed ω and the translational speed of its center of rotation v_r (where the r subscript is to indicate rolling) are related by $v_\mathrm{r} = \omega R$, with R the relevant radius of our object. If the object's center of rotation moves faster than v_r, the rotation can't 'keep up', and the object slides

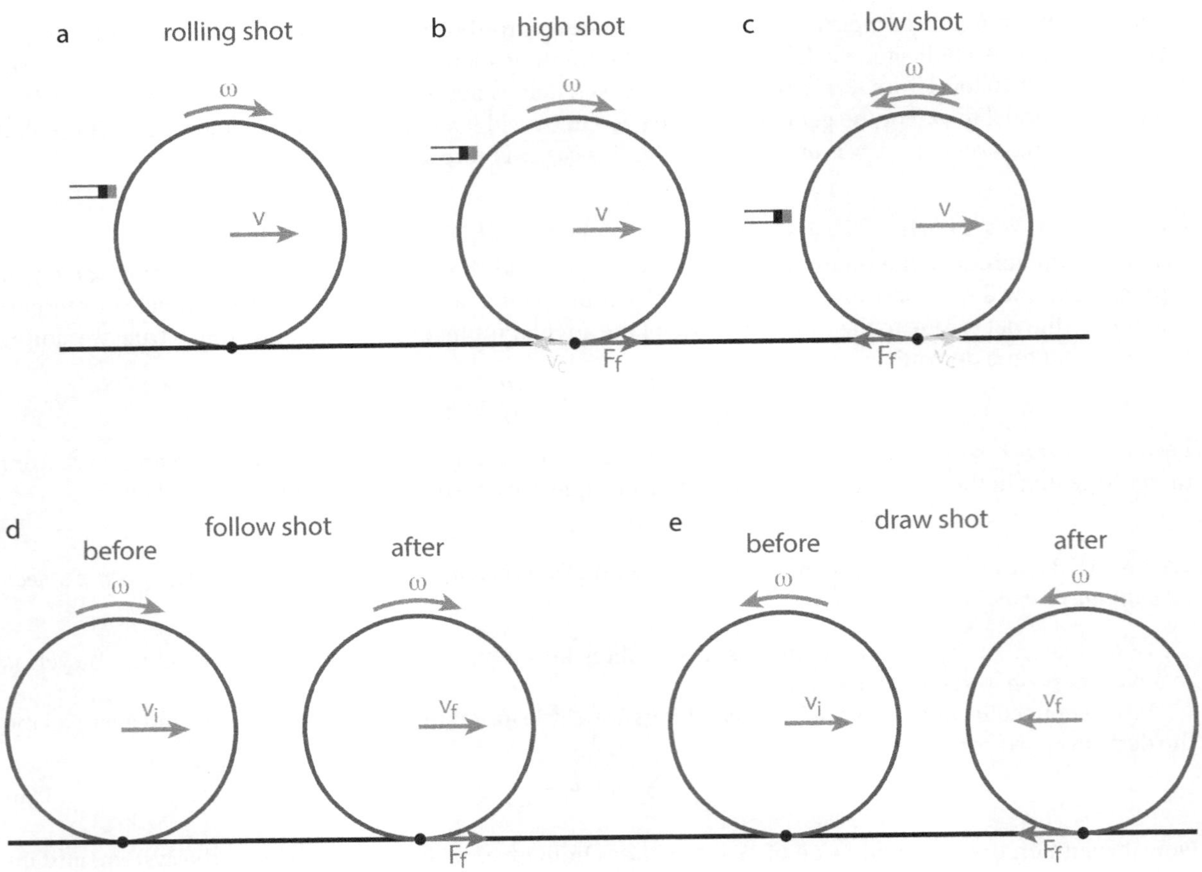

Figure 5.1: Five types of billiard shots. (a-c) The type of motion depends on where the cue hits the ball. (a) If the cue hits the ball at exactly $7R/5$ above the table, the ball will exhibit pure rolling motion, $\omega = vR$. (b) If the cue hits the ball above the critical spot, it will rotate faster than translate ($\omega > vR$) and exhibit a slipping rotation. Friction will slow down the rotation until rolling motion is achieved. (c) If the cue hits the ball below the critical spot, it will translate faster than rotate ($\omega < vR$) and initially slide, until friction both slows down the translational speed and accelerates the rotational speed to the point where rolling motion is achieved. Note that the rotational motion may even be retrograde, i.e., backwards compared to the translational motion. (d-e) Behavior of the incident billiard ball before and after collision with a stationary ball of equal mass. Since the collision is elastic, all linear momentum is transferred to the other ball. If the incident ball was initially rolling, immediately after the collision it will continue rotating with complete slipping. Friction then causes the ball to pick up linear speed again, with a direction depending on the direction of the rotational motion, resulting in a follow (d) or draw (e) shot.

over the surface. We call this type of motion *slipping*. Due to friction, objects undergoing slipping motion typically quickly slow down to v_r, at which point they roll without slipping.

Suppose we started our object with a velocity v_0. If there is no rotation, the only force changing its velocity is the constant frictional force $F_{\text{friction}} = \mu_k F_N = \mu_k mg$, with m the mass of the object (equation 2.13). The constant force results in a linear decrease in the translational speed (see section 2.3): $v(t) = v_0 - \mu_k g t$. However, if our object can roll, there is a second contribution to the motion, due to the torque $\tau_{\text{friction}} = F_{\text{friction}} R$ of the frictional force. Using the rotational analog of Newton's second law, equation (5.12) (or writing $L = I\omega$ and using equation (5.25)), we get an equation of motion for the rotational velocity:

$$I\alpha = I\frac{d\omega}{dt} = \tau_{\text{friction}} = F_{\text{friction}} R. \tag{5.27}$$

Integrating equation (5.27) with initial condition $\omega(t=0) = 0$ we get $\omega(t) = \mu_k mgRt/I$. While the object undergoes slipping motion, the translational speed thus linearly decreases with time, whereas the rotational speed linearly increases. To find the time and velocity at which the object enters a purely rolling motion, we simply equate $v(t)$ with $\omega(t)R$, which gives

$$t_r = \frac{v_0}{\mu_k g(1 + mR^2/I)}, \tag{5.28}$$

$$v_r = \frac{v_0}{1 + I/(mR^2)}. \tag{5.29}$$

Note that the time t_r until fully rolling motion is achieved scales inversely with the friction coefficient, but the final rolling speed v_r is independent of the frictional force. The rolling speed does depend on the moment of inertia of your object - for a hollow cylinder it's $v_r = \frac{1}{2}v_0$, whereas for a solid cylinder it's $v_r = \frac{2}{3}v_0$. Once the object is rolling, its surface no longer moves with respect to the surface that it's rolling on (as its instantaneous speed at the point of touching is zero). Consequently, the frictional force is much reduced, and the object can roll a large distance before it stops; in fact, the main force slowing it down once it is rolling is drag with the ambient air, which we could safely ignore when (kinetic) friction was still in the picture.

5.8.1 WORKED EXAMPLE: A CYLINDER ROLLING DOWN A SLOPE

A massive cylinder with mass m and radius R rolls without slipping down a plane inclined at an angle θ. The coefficient of (static) friction between the cylinder and the plane is μ. Find the linear acceleration of the cylinder.

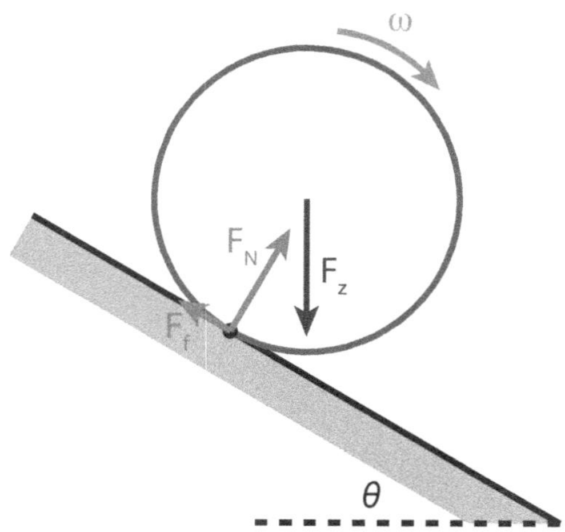

Figure 5.2: Free body diagram of a cylinder rolling down a plane.

SOLUTION

There are at least three ways to tackle this problem. For all three, it helps (as always) to make a sketch, indicating the relevant forces - see figure 5.2.

- *Method 1: Forces and torques.* Let the friction force F_f be positive in the direction up the plane. Then we have:

$$\begin{aligned} F = ma \quad &\Rightarrow \quad mg\sin\theta - F_f = ma \\ \tau = I\alpha \quad &\Rightarrow \quad F_f R = \frac{1}{2} mR^2 \alpha \\ \text{no slipping} \quad &\Rightarrow \quad a = \alpha R. \end{aligned}$$

The last two equations give $F_f = \frac{1}{2}ma$. Plugging this into the first equation gives

$$a = \frac{g\sin\theta}{1+\frac{1}{2}} = \frac{2}{3} g\sin\theta.$$

- *Method 2: Energy.* The total energy of the system is given by

$$E_{tot} = K + V = \frac{1}{2}mv^2 + \frac{1}{2}I\omega^2 + mgh.$$

If the cylinder rolls down the slope without slipping, its angular and linear velocities are related through $v = \omega R$. Also, if it moves a distance Δx, its height decreases by $\Delta x \cdot \sin\theta$. Conservation of energy then gives:

$$\begin{aligned} 0 = \frac{dE_{tot}}{dt} &= \frac{d}{dt}\left[\frac{1}{2}mv^2 + \frac{1}{2}I\left(\frac{v}{R}\right)^2 - mgx\sin\theta\right] \\ &= mv\dot{v} + I\frac{v\dot{v}}{R^2} - mgv\sin\theta \\ &= \left[a + \frac{1}{2}a - g\sin\theta\right] mv, \end{aligned}$$

where we used $I = \frac{1}{2}mR^2$ for a massive cylinder in the last line. The linear acceleration a is thus given by $a = \frac{2}{3}g\sin\theta$.

- *Method 3: Rotational version of Newton's second law.* At a given point in time, we can apply the rotational version of Newton's second law to rotations about the point where the cylinder touches the surface (as the cylinder is rolling without slipping, this is the only motion at that point). Of the three forces in the system, two act at that point, so they have no lever arm. Only gravity has a nonzero lever arm of length $R\sin\theta$, leading to a torque given by $\tau_z = mgR\sin\theta$. By the rotational version of Newton's second law, we have $\tau = I\alpha$, where I is the moment of inertia about the pivot. Applying the parallel-axis theorem, we find $I = I_{cm} + md^2 = \frac{3}{2}mR^2$ in this case, so we get an angular acceleration of

$$\alpha = \frac{\tau_z}{I} = \frac{mgR\sin\theta}{\frac{3}{2}mR^2} = \frac{2g}{3R}\sin\theta.$$

The linear acceleration of the center of the cylinder due to the 'rotation' about this pivot is given by $a = R\alpha = \frac{2}{3}g\sin\theta$.

5.9 PRECESSION AND NUTATION

The action of a torque causes a change in angular momentum, as expressed by equation 5.25). A special case arises when the torque is perpendicular to the angular momentum: in that case the change affects only the direction of the angular momentum vector, not its magnitude. Since the torque is given by the cross product of the arm and the force, this case arises when the angular momentum is parallel to either arm or force, or more generally, lies in the plane spanned by the force and arm. As a result, the angular momentum vector may start rotating about a fixed axis, a process known as *precession.* Due to the action of a second force (with associated torque), the angle between the angular momentum vector and the fixed axis (which we'll call the z-axis) may also change, a process known as *nutation.*

The simplest example of a precessing system is that of a rotationally symmetric top, spinning about an axis that is not the vertical z-axis, see figure 5.3. In this case, the arm of gravitational force (pointing from the

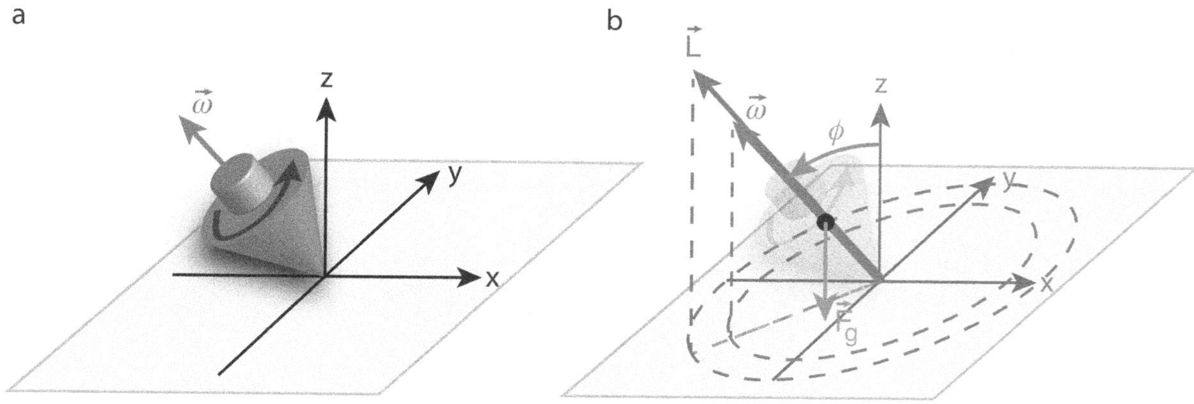

Figure 5.3: Precession of a spinning top. The top rotates about an axis which makes an angle ϕ with the vertical (z) axis. The angular momentum of the top is parallel to the rotation vector. The arm between the pivot (where the top touches the supporting surface) and the center of mass is parallel to the angular momentum as well. Consequently, the torque of the gravitational force (which as always is pointed downward from the center of mass) is perpendicular to the angular momentum, and causes it to *precess* about the z-axis with a precession frequency ω_{p}.

pivot at the origin to the center of mass of the top) is parallel to the angular momentum, which itself is parallel to the rotation vector, as $\boldsymbol{L} = I\boldsymbol{\omega}$. The torque of gravity is thus perpendicular to the angular momentum. If we call the angle between $\boldsymbol{L}$ and the z-axis ϕ, and work in cylindrical coordinates (ρ, θ, z), we can write $\boldsymbol{L} = L_{xy}\hat{\boldsymbol{\rho}} + L_z\hat{\boldsymbol{z}}$, where $L_{xy} = L\sin\phi$ is the projection of $\boldsymbol{L}$ on the xy-plane, and $\hat{\boldsymbol{\rho}}$ is the radial unit vector in the xy-plane (i.e., $\hat{\boldsymbol{\rho}} = \cos\theta\hat{\boldsymbol{x}} + \sin\theta\hat{\boldsymbol{y}}$). The gravitational torque is then given by:

$$\boldsymbol{\tau}_{\mathrm{z}} = \boldsymbol{r} \times \boldsymbol{F}_{\mathrm{z}} = mgr\sin\phi\hat{\boldsymbol{\theta}}, \tag{5.30}$$

where $\boldsymbol{r}$ is the arm pointing from the origin to the center of mass, r its length, and $\hat{\boldsymbol{\theta}}$ the angular unit vector in the xy-plane (i.e., $\hat{\boldsymbol{\theta}} = -\sin\theta\hat{\boldsymbol{x}} + \cos\theta\hat{\boldsymbol{y}}$). For the time derivative of $\boldsymbol{L}$, we get:

$$\frac{\mathrm{d}\boldsymbol{L}}{\mathrm{d}t} = \frac{\mathrm{d}L_{xy}}{\mathrm{d}t}\hat{\boldsymbol{\rho}} + L_{xy}\dot{\theta}\hat{\boldsymbol{\theta}} + \frac{\mathrm{d}L_z}{\mathrm{d}t}\hat{\boldsymbol{z}}, \tag{5.31}$$

where we used (see equation A.8)

$$\frac{\mathrm{d}\hat{\boldsymbol{\rho}}}{\mathrm{d}t} = \frac{\mathrm{d}\hat{\boldsymbol{\rho}}}{\mathrm{d}\theta}\frac{\mathrm{d}\theta}{\mathrm{d}t} = \dot{\theta}\hat{\boldsymbol{\theta}}. \tag{5.32}$$

Equating (5.30) and (5.31), we find that $\mathrm{d}L_{xy}/\mathrm{d}t = \mathrm{d}L_z/\mathrm{d}t = 0$, and

$$\omega_{\mathrm{p}} \equiv \frac{\mathrm{d}\theta}{\mathrm{d}t} = \frac{mgr\sin\phi}{L_{xy}} = \frac{mgr}{I\omega}. \tag{5.33}$$

Equation (5.33) gives us the frequency of the precession about the z-axis. The associated precession rotation vector is given by $\boldsymbol{\omega}_{\mathrm{p}} = \omega_{\mathrm{p}}\hat{\boldsymbol{z}}$.

Where precession, in terms of the angles used in this section, represents a change in θ, nutation is associated with a change in the tilt angle ϕ. It comes about in several cases, most commonly due to the action of a second force, and often results in a periodic motion (hence the name nutation, 'nodding'). One example is the change in the Earth's axis. As the axis is not perpendicular to the plane of Earth's orbit (it presently makes an angle of about 23.4°), gravity from the sun exerts a net torque on the Earth's rotation axis, causing it to precess with a period of about 25000 years. Consequently, the current polar star will only remain the polar star for a couple of centuries. Due to the torques exerted by the moon and the other planets, the Earth's axis also nutates - with amplitudes from arcseconds to a few degrees, and periods that range widely, from 18.6 years due to the gravitational pull of the moon up to millennia due to other effects.

5.10 PROBLEMS

5.1 Figure 5.4 shows two hand powered drills.

(a) With which of the two drills shown in figure 5.4 will you be able to exert a greater torque on the drill bit?

(b) With which of the two drills shown in figure 5.4 will you be able to exert a greater rotational speed?

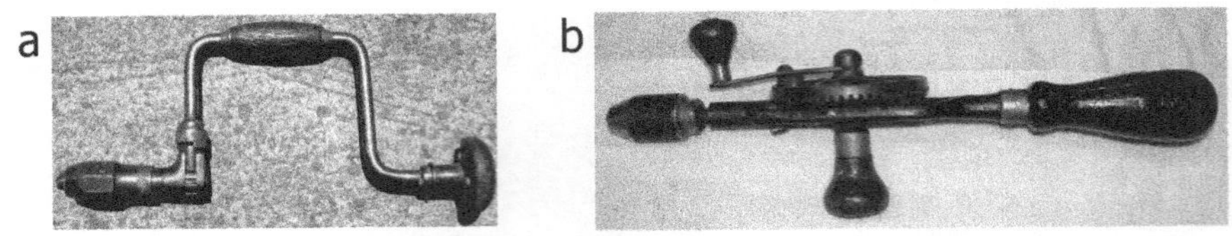

Figure 5.4: Two hand-powered drills. (a) A brace [16], CC BY-SA 3.0. (b) An egg-beater drill [17].

5.2 An Atwood's machine consists of two masses m_1 and m_2, connected by a string that passes over a pulley. If the pulley is a disk of radius R and mass M, find the acceleration of the masses.

5.3 You hold a uniform pen with a mass of 25.0 g horizontal with your thumb pushing down on one end and your index finger pushing upward 2.0 cm from your thumb. The pen is 14 cm long.

(a) Which of your fingers exerts the greater force?

(b) Find the two forces.

5.4 A wagon wheel is constructed as shown in the figure below. The radius of the wheel is R. Each of the spokes that lie along the diameter has a mass m, and the rim has mass M (you may assume the thickness of the rim and spokes are negligible compared to the radius R).

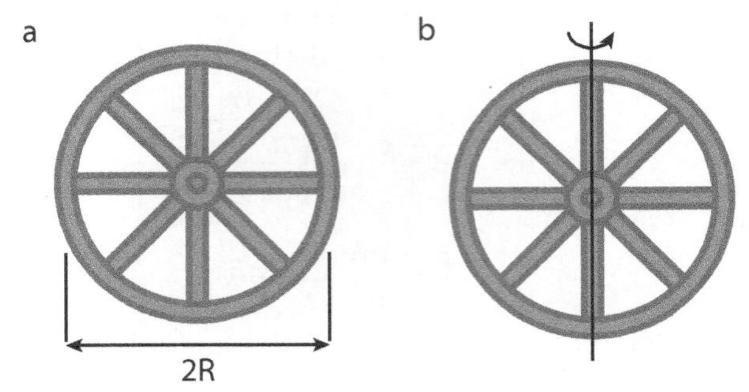

(a) What is the moment of inertia of the wheel about an axis through the center, perpendicular to the plane of the wheel (figure a)?

(b) For the same wheel, what is the moment of inertia for an axis through the center and two of the spokes, in the plane of the wheel (figure b)?

5.5 A sphere with radius $R = 0.200$ m has a density ρ that decreases with distance r from the center of the sphere, according to $\rho(r) = a - br$, where $a = 1.00 \cdot 10^3$ kg/m^3 and $b = 4.00 \cdot 10^3$ kg/m^4.

(a) Calculate the total mass of the sphere.

(b) Calculate the moment of inertia of the sphere for an axis through its center.

5.6 A 0.10 kg yo-yo has an outer radius R that is 5.0 times greater than the radius r of its axle. The yo-yo is in equilibrium if a mass m is suspended from its outer edge, as shown in figure 5.5. Find the tensions T_1 and T_2 in the two strings, and the mass m.

5.7 A table consists of three pieces: a tabletop which is a circular disk of radius R, thickness $d = R/50$, and mass M; a single leg that supports the tabletop in its center and consists of a hollow cylinder of height $h = R/2$ and mass $m = M/5$, and a foot, which consists of a solid cylinder of radius $R/3$, mass M and height $h = R/4$.

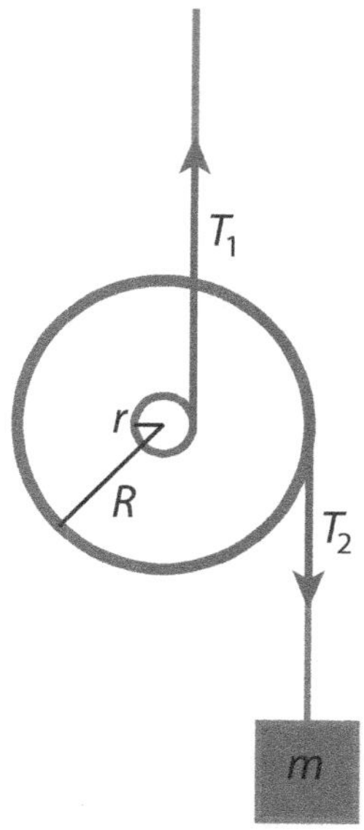

Figure 5.5: A yo-yo with a mass suspended from its outer edge.

(a) Find the position of the center of mass of this table.

(b) With what force should you push down on the edge of the table to make it tip over?

(c) A stone of mass $m = M/2$ is placed on the table. How far out from the center can it be positioned before the table tips over? You may approximate the stone as a point mass.

5.8 A set-square protractor made of plastic with density $\rho = 1.2\,\mathrm{g/cm^3}$, thickness 2.0 mm and sides of 10 cm is placed in a coordinate system as shown in the figure. The z-axis (not shown) is coming out of the paper.

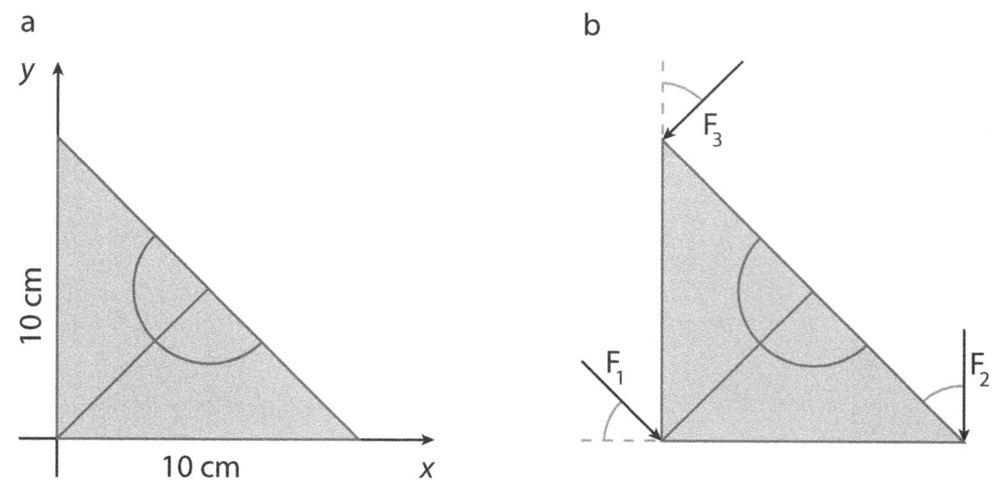

(a) Find the position of the center of mass of the protractor.

(b) Find the moment of inertia of the protractor when rotating it along the y-axis.

(c) Find the moment of inertia of the protractor when rotating it along the z-axis (i.e., the axis through the origin and perpendicular to the xy-plane).

(d) Find the moment of inertia of the protractor when rotating it along an axis through its center of mass, and parallel to the z-axis (i.e., perpendicular to the xy-plane).

(e) You pick up the protractor and balance it on your finger in its center of mass. You then tap one of the points with a small force F, directed in the plane of the protractor. For which of the three forces shown in figure b is the rotational speed the protractor gets the largest? (As always, explain your answer; the magnitude of all three forces is the same, the three indicated angles are all 45°).

(f) You bring in three friends, who now push on the protractor along the directions indicated by F_1, F_2 and F_3 in the figure. If F_1 and F_2 are both 1.0 N, how large must F_3 be so that the protractor does not move?

5.9 In this problem, we consider an object which consists of a frame with four balls, as depicted in the figure below. Each of the balls has a radius $R = 5.0$ cm. Their masses are $m_A = 3.0$ kg, $m_B = 3.0$ kg, $m_C = 2.0$ kg and $m_D = 1.0$ kg. The rods of the frame (which run from the edge of one ball to the edge of the next) have lengths $L_1 = 20$ cm and $L_2 = 30$ cm and a linear density $\lambda = 10$ g/cm. (The distance between the centers of balls A and B is thus 30 cm, and between the centers of balls B and C is 40 cm). The thickness of the rods is negligible. We use the xyz frame indicated in the figure, with a z axis coming out of the paper in the origin.

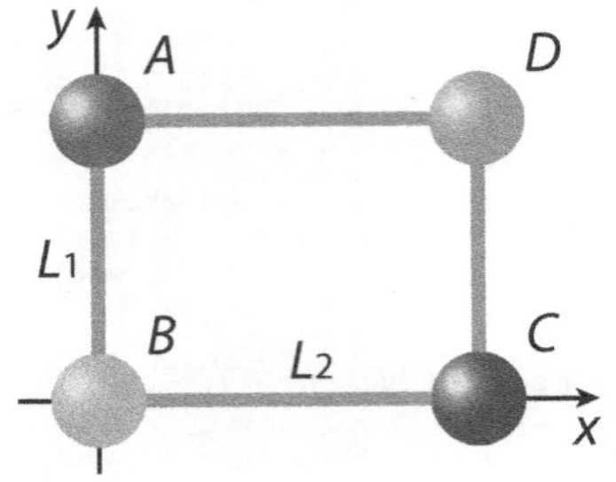

(a) Find the position of the center of mass of the object.

(b) Find the moment of inertia of the object with respect to rotations about the y-axis.

(c) We apply a torque of 10 Nm to set the object spinning about the y-axis. Calculate the magnitude of the resulting angular velocity.

(d) Find the moment of inertia of the object with respect to rotations about an axis through its center of mass, and parallel to the y-axis.

(e) Find the moment of inertia of the object with respect to rotations about an axis through its center of mass, and parallel to the z-axis.

5.10 Three uniform bars with linear density (i.e., mass per unit length) λ are welded together into the shape of an isosceles triangle with sides of length $5L$, $5L$, and $6L$ (figure a).

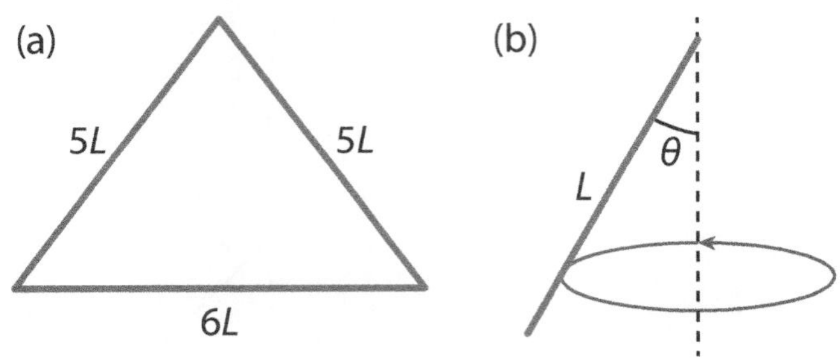

The moment of inertia of a rod with mass m and length L about an axis that goes through one of its end points at an angle θ with the rod itself (figure b) is given by

$$I_{\text{bar}} = \frac{1}{3} m L^2 \sin^2\theta.$$

(a) Find the position of the center of mass of the triangle.

(b) Find the moment of inertia of the triangle about its axis of symmetry. To do so, you may use the given expression for the moment of inertia of a bar at an angle.

(c) Find the moment of inertia of the triangle about its longest side.

(d) Find the moment of inertia of the triangle about the axis through its center of mass, perpendicular to the triangle's plane. Use the theorems proved in this section to do so.

(e) Are rotations of this triangle about its symmetry axis stable?

(f) [Challenging] Derive the given expression for the moment of inertia of the rod under an angle.

5.11 We consider the same setting as in the worked example in section 5.8. A massive cylinder with mass m and radius R rolls without slipping down a plane inclined at an angle θ (see figure a). The coefficient of (static) friction between the cylinder and the plane is μ.

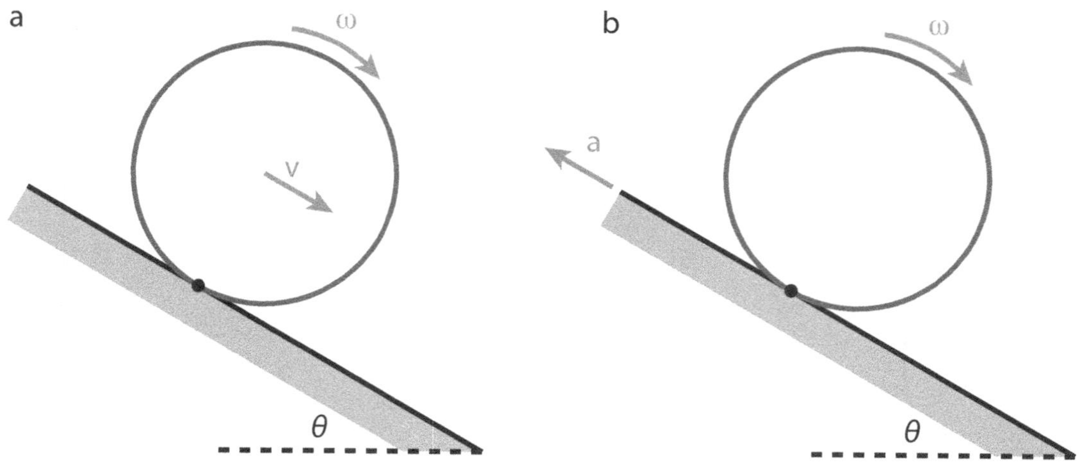

(a) Find the largest angle θ for which the cylinder doesn't slip.

(b) Suppose now that we accelerate the plane upwards (along its direction) with acceleration a, see figure b. For what value of a does the center of mass of the cylinder not move? (You may assume that the cylinder still isn't slipping).

5.12 We consider a wheel with mass m and radius R which rolls without slipping on a flat (problems a and b) and an inclined (problems c-i) plane.

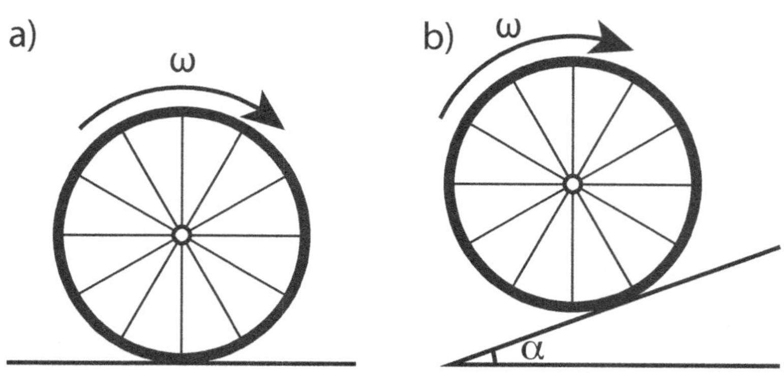

(a) Draw a free-body diagram of the wheel rolling on a horizontal surface (figure a). Also indicate the (linear) velocity vector $\boldsymbol{v}$ of the wheel.

(b) How much work is done by the friction force on the wheel? What effect does this work have on the linear and angular velocity of the wheel?

We now consider the case that the wheel rolls up an inclined plane (angle α, see figure b) with initial velocity v_0.

(c) Again draw a free-body diagram of the wheel and indicate the linear velocity $\boldsymbol{v}$.

(d) Determine the equation of motion for both the linear and the rotational velocity of the wheel (i.e. apply Newton's second law to the translational and rotational motion ot the wheel). For this problem, you can describe the wheel as a cylindrical ring (and thus ignore the mass of the spokes).

(e) Solve the equations of motion from (12d) to obtain the linear acceleration of the wheel.

(f) From the equations of motion, determine the magnitude and the direction of the frictional force acting on the wheel.

(g) Determine the total energy of the wheel.

(h) Take the derivative of the total energy of problem (12g) to again arrive at an equation of motion for the wheel.

(i) Is the energy of the wheel rolling up the incline conserved? Does the frictional force do any work in this case?

5.13 We consider a solid cone of uniform density ρ, height h and base radius R.

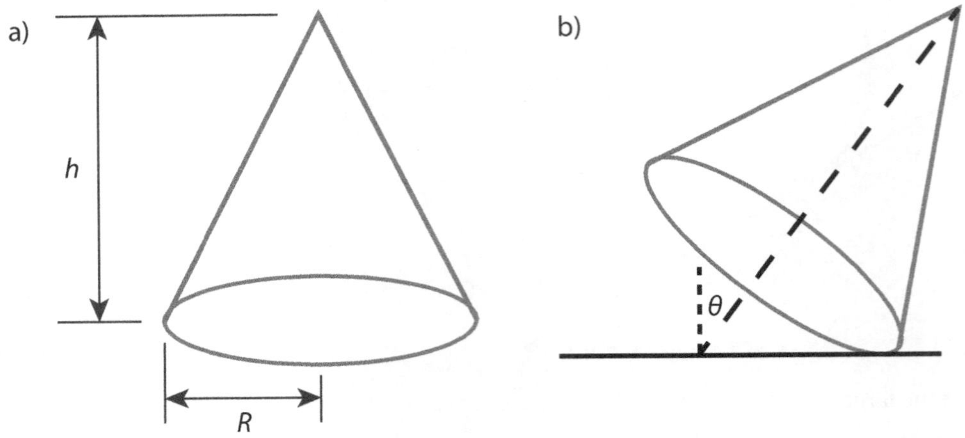

Figure 5.6: Sketch of the cone. (a) Dimensions. (b) Tilt angle θ.

(a) Taking the origin to be at the center of the base, and the symmetry axis along the z (vertical) axis, show that the center of mass of the cone is located at $(0,0,h/4)$. *Hint*: consider the cone as a stack of disks and pick your origin carefully.

(b) Over what maximum angle can you tilt the cone such that it just doesn't fall over?

(c) Qualitatively sketch the potential energy landscape $U(\theta)$ of the cone as a function of this tilt angle θ. Indicate the stationary points and their stability. Assume the cone is narrow, i.e., the radius of the base is less than a quarter of the height.

(d) Going back to the upright position, show that the moment of inertia of the cone around its axis of symmetry is given by $I = (3/10)MR^2$, where M is the total mass of the cone.

(e) Consider a cone which is 100 cm high, has a base radius of 20 cm, and a mass density of 700 kg/m^3. If the cone is standing with its base on a perfectly slippery surface (no friction), and you want to set it spinning at 1 revolution per minute, starting from rest, how much work do you have to do?

5.14 A hockey stick of mass m_s, length L, and moment of inertia (with respect to its center of mass) I_0 is at rest on the ice (which we assume to be frictionless). A puck with mass m_p hits the stick a distance D from the middle (i.e., center of mass) of the stick. Before the collision, the puck was moving with speed v_0 in a direction perpendicular to the stick, as indicated in the figure. The collision is completely inelastic, and the puck remains attached to the stick after the collision.

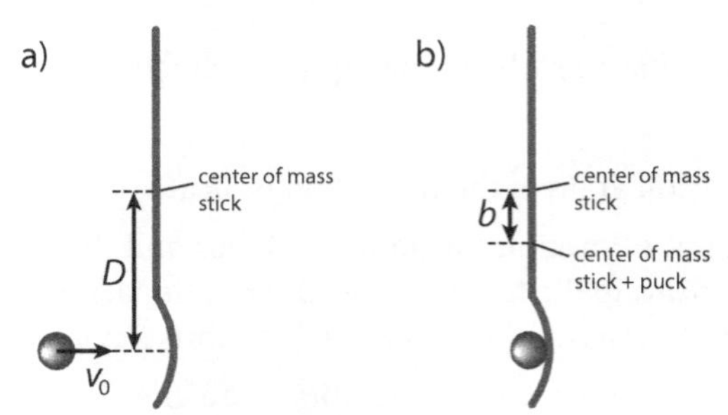

(a) Find the speed v_f of the center of mass of the stick + puck combination after the collision, in terms of v_0, m_p, m_s, and L.

(b) After the collision, the stick and puck will rotate about their combined center of mass, which lies a distance b from the center of mass of the stick. Find b.

(c) What is the angular momentum L_{cm} of the system before the collision, with respect to the center of mass of the final system? Use your answer at (b) to eliminate the value of b from your answer.

(d) What is the angular velocity ω of the stick + puck combination after the collision? Your answer should be independent of the variable b.

(e) Are the kinetic energy, linear momentum, and angular momentum conserved in this collision?

5.15 We consider the same situation as in problem 5.14, but now the collision is fully elastic. Find the velocity of the ball after the collision, as well as the translational and rotational velocity of the stick after the collision. *Hint*: You have to solve for three unknowns, so you need three equation - what are they?

5.16 In Greek mythology Sisyphus by life was the king of Ephyra (Corinth). After he died, he was punished for his self-aggrandizing craftiness and deceitfulness in Tartarus (Greek hell) by being forced to roll an immense boulder up a hill only for it to roll down when it nears the top, repeating this action for all eternity. We take the original rock to have a diameter of 1.00 m and a mass of 2000 kg; the hill is 1.00 km high and has a slope of 30°.

(a) Suppose Sisyphus somehow managed to replace this rock with one made of aerogel with a mass of only 1.0 kg covered by a thin layer of rock which also has a mass of 1.0 kg. To hide his deceit, he still walks uphill slowly, and kicks the rock when it 'slips' down again. What speed must Sisyphus give the fake rock such that it has the same speed at the bottom as the real rock would have? You may assume that both rocks roll all the way down without slipping.

(b) Hades, being not amused with Sisyphus' trickery in part (a), decides to repay him using his own measures. He sends Sisyphus off to Helheim (Norse hell), strapped to the original rock, which he kicks so hard that it reaches escape velocity (11.2 km/s) while also spinning at a rate of 1.0 revolutions per second. How much energy did Hades put into the rock (you may ignore Sisyphus' mass, he's a ghost anyway)?

(c) Hel, the ruler of Helheim, doesn't want him either, so she picks up the capstone of a dolmen, with a mass ten times that of Sisyphus' rock, and hurls it towards the oncoming Sisyphus. As it happens, Hel throws just hard enough that after the collision, Sisyphus (now stuck between the two rocks) is spinning in place. How fast did Hel throw her stone?

(d) At what angular velocities do Sisyphus and his two rocks spin after the collision?

(e) Dante finally finds a place for Sisyphus in Inferno (Christian hell). To get him there, he first needs to remove the capstone. Taking pity on Sisyphus, Dante wishes to do so in a way that the original rock stops spinning. He stands on Sisyphus' rock, pushing on the capstone. What torque must he exert on that stone to stop Sisyphus' rock?

(f) Dante then ties a 100 km long rope to the rock, which at the other end is connected to Lucifer (down at the center of Inferno). Lucifer then hauls in the rope, with a speed of 1.0 m/s. Unfortunately for Sisyphus, the removal of the capstone, although it has stopped him from spinning, has resulted in him getting a small linear velocity of 1.0 cm/s in the direction perpendicular to the rope. When Lucifer has hauled in enough rope to put Sisyphus in the fourth circle of Inferno at about 5 km from himself, what is the angular velocity of Sisyphus' rock?

5.17 Two children, with masses $m_1 = 10$ kg and $m_2 = 10$ kg sit on a simple merry-go-round that consists of a massive disk with a mass of 100 kg and a radius of 2.0 m. The disk can rotate freely about its center, and is doing so at a frequency ω_0 of 5.0 revolutions per minute. A third child with mass $m_3 = 10$ kg runs towards the merry-go-round with a speed v_0 of 1.0 m/s, at an angle of 30° with the tangent to the merry-go-round's edge (see figure). Once it reaches the merry-go-round, the child jumps on it, and continues rotating with the other two. Find the resulting rotational velocity of the merry-go-round with the three children.

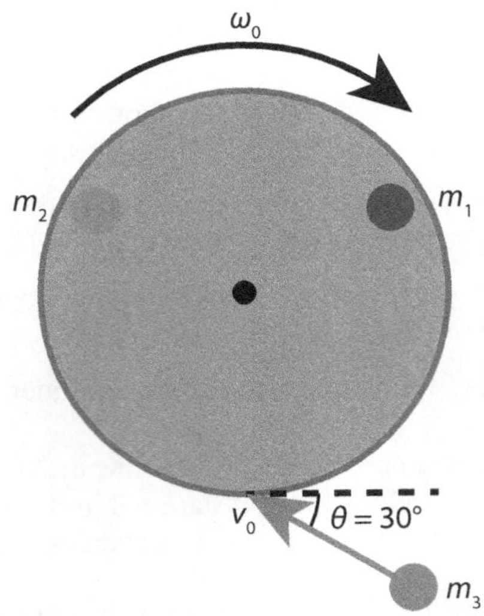
ω0
m2
m1
v0
θ = 30°
m3

6

GENERAL PLANAR MOTION

Although Newton's laws of motion, the various force laws, and the three conservation laws we have derived, are all valid in three dimensions, we have so far restricted our study of motion almost exclusively to two special cases: linear motion in one dimension, and rotational motion in a plane, where the radius of the rotation is constant. Although for the second case we do need two directions to describe it, the motion itself is constricted to a circle, and thus essentially one-dimensional. In this section, we'll look at general motion in a plane - which turns out to capture a large number of important nontrivial cases.

6.1 PROJECTILE MOTION

The simplest case of two-dimensional motion occurs when a particle experiences a force only in one direction. The prime example of this case is the motion of a projectile in Earth's (or any other planet's) gravitational field as locally described by Galilean gravity (equation 2.8): $\boldsymbol{F} = m\boldsymbol{g}$. Once a projectile has been fired with a certain initial velocity $\boldsymbol{v}_0$, we can find its trajectory by solving the equation of motion that follows from Newton's second law: $m\boldsymbol{g} = m\ddot{\boldsymbol{r}}$. We can decompose $\boldsymbol{r}$ and $\boldsymbol{v}_0$ in horizontal (x) and vertical (z) components; each of them has its own one-dimensional equation of motion, which we already solved in section 2.3. The horizontal component experiences no force and thus executes a simple linear motion with uniform velocity $v_0 \cos\theta_0$, where $\theta_0 = \arccos(\boldsymbol{v}_0 \cdot \hat{\boldsymbol{x}})/v_0$ is the angle with the horizontal under which the projectile was fired and $v_0 = |\boldsymbol{v}_0|$ the initial speed. Likewise, because the acceleration due to gravitation is constant, our projectile will execute a uniformly accelerated motion in the vertical direction with initial velocity $v_0 \sin\theta_0$. If the projectile's initial position is (x_0, z_0), its motion is thus described by:

$$\boldsymbol{r}(t) = \begin{pmatrix} x(t) \\ z(t) \end{pmatrix} = \begin{pmatrix} x_0 \\ z_0 \end{pmatrix} + v_0 \begin{pmatrix} \cos\theta_0 \\ \sin\theta_0 \end{pmatrix} t - \begin{pmatrix} 0 \\ g \end{pmatrix} \frac{1}{2} t^2. \tag{6.1}$$

6.2 GENERAL PLANAR MOTION IN POLAR COORDINATES

Although in principle all planar motion can be described in Cartesian coordinates, they are not always the easiest choice. Take, for example, a central force field (a force field whose magnitude only depends on the distance to the origin, and points in the radial direction), as we'll study in the next section. For such a force field polar coordinates are a more natural choice than Cartesians. However, polar coordinates do carry a few subtleties not present in the Cartesian system, because the direction of the axes depends on position. We will therefore first derive the relevant expressions for the position, velocity and acceleration vector, as well as the components of the force vector, in polar coordinates for the general case.

As we already know (see appendix A.2), the position vector $\vec{r} = x\hat{\boldsymbol{x}} + y\hat{\boldsymbol{y}}$ has a particularly simple expression in polar coordinates: $\boldsymbol{r} = r\hat{\boldsymbol{r}}$, where $r = \sqrt{x^2 + y^2}$. To find the velocity and acceleration vectors in polar coordinates, we take time derivatives of $\boldsymbol{r}$. Note that because the orientation of the polar basis vectors depends on the position in space, the time derivative acts on both the distance to the origin r and the basis vector $\hat{\boldsymbol{r}}$. Because the two polar basis vectors are each other's derivatives with respect to θ (see equation A.8), we find for their time derivatives:

$$\frac{\mathrm{d}\hat{\boldsymbol{r}}}{\mathrm{d}t} = \frac{\mathrm{d}\hat{\boldsymbol{r}}}{\mathrm{d}\theta}\frac{\mathrm{d}\theta}{\mathrm{d}t} = \dot{\theta}\hat{\boldsymbol{\theta}}, \qquad \frac{\mathrm{d}\hat{\boldsymbol{\theta}}}{\mathrm{d}t} = \frac{\mathrm{d}\hat{\boldsymbol{\theta}}}{\mathrm{d}\theta}\frac{\mathrm{d}\theta}{\mathrm{d}t} = -\dot{\theta}\hat{\boldsymbol{r}}. \tag{6.2}$$

For the velocity and acceleration vectors we then find:

$$\boldsymbol{v} = \frac{\mathrm{d}\boldsymbol{r}}{\mathrm{d}t} = \dot{r}\hat{\boldsymbol{r}} + r\dot{\theta}\hat{\boldsymbol{\theta}}, \tag{6.3}$$

$$\boldsymbol{a} = \frac{\mathrm{d}\boldsymbol{v}}{\mathrm{d}t} = (\ddot{r} - r\dot{\theta}^2)\hat{\boldsymbol{r}} + (r\ddot{\theta} + 2\dot{r}\dot{\theta})\hat{\boldsymbol{\theta}}. \tag{6.4}$$

Note that equations (5.3) and (5.6) are the special cases of equations (6.3) and (6.4) for which both the radius r and the angular velocity $\omega = \dot{\theta}$ are constant.

Using equation (6.4) for $\ddot{\boldsymbol{r}}$ in Newton's second law, we get an expression decomposing the net force $\boldsymbol{F}$ into a radial and an angular part, each of which consists of two terms:

$$\boldsymbol{F} = m\ddot{\boldsymbol{r}} = F_r\hat{\boldsymbol{r}} + F_\theta\hat{\boldsymbol{\theta}} \tag{6.5}$$

$$F_r = m(\ddot{r} - r\dot{\theta}^2) \tag{6.6}$$

$$F_\theta = m(r\ddot{\theta} + 2\dot{r}\dot{\theta}) \tag{6.7}$$

The two terms in F_r are readily identified as the radial acceleration $\ddot{r}$ (acting along the line through the origin) and the centripetal force (which causes objects to rotate around the origin, see equation (5.10)). The first term $r\ddot{\theta}$ in F_θ is the tangential acceleration α of a rotating object whose angular velocity is changing (equation (5.8)). The last term in F_θ we have not encountered before; it is known as the *Coriolis force*

$$\mathbf{F}_{\mathrm{Cor}} = 2m\dot{r}\dot{\theta}\hat{\boldsymbol{\theta}}, \tag{6.8}$$

and is associated with a velocity in both the radial and the angular direction. It is fairly weak on everyday length scales, but gets large on global length scales. In particular, if you move over the surface of the Earth (necessarily with a nonzero angular component of your velocity), it tends to deflect you from a straight path. On the Northern hemisphere, if you move horizontally, it tends to push you to the right; it also pushes you west when going up, and east when going down. Coriolis forces are responsible for the rotational movement of air around high and low pressure zones, causing respectively clockwise and counterclockwise currents around them on the Northern hemisphere (see figure 6.1). We'll encounter the Coriolis force again in the more general three-dimensional setting in section 7.2.

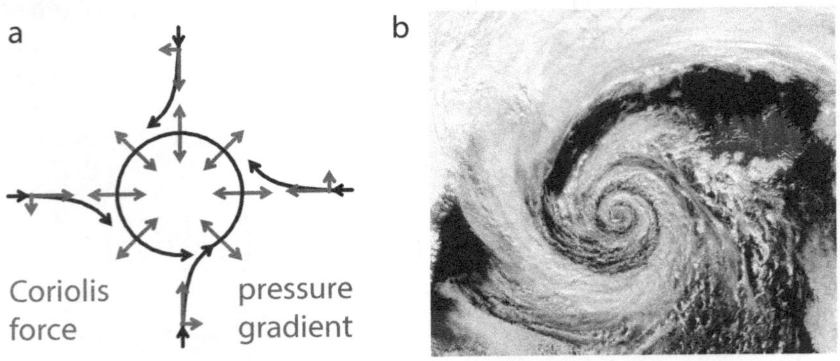

Figure 6.1: The Coriolis force causes clockwise and counterclockwise currents around high and low pressure zones on the Northern hemisphere. (a) Pressure gradient (blue), Coriolis force (red) and resulting air flow (black) around a low pressure zone. (b) Typical satellite picture of a low-pressure zone and associated winds over Iceland. Picture by NASA's Aqua/MODIS satellite [20].

6.3 Motion under the action of a central force

A *central force* is a force that points along the (positive or negative) radial direction $\hat{\boldsymbol{r}}$, and whose magnitude depends only on the distance r to the origin - so $\boldsymbol{F}(\boldsymbol{r}) = F(r)\hat{\boldsymbol{r}}$. Central forces can be defined in both two and three dimensions, with the three-dimensional concept of the radial distance (to the origin) and direction (direction of increasing r) completely analogous to the two-dimensional case. Two important examples of central forces are (general) Newtonian gravity (2.9) and the Coulomb force (2.10) between two charged objects. Although these forces are three-dimensional examples, discussing them here is appropriate, as the following theorem shows.

Theorem 6.1. *The motion of a particle under the action of a central force takes place in a plane.*

Proof. We first note that a central force can exert no torque on an object: $\boldsymbol{\tau} = \boldsymbol{r} \times \boldsymbol{F} = F(r)(\boldsymbol{r} \times \hat{\boldsymbol{r}}) = 0$. Consequently, under the action of a central force, angular momentum is conserved. Moreover, we have $\boldsymbol{r} \cdot \boldsymbol{L} = \boldsymbol{r} \cdot (\boldsymbol{r} \times \boldsymbol{p}) = 0$ and $\boldsymbol{v} \cdot \boldsymbol{L} = \boldsymbol{v} \cdot (\boldsymbol{r} \times m\boldsymbol{v}) = 0$. Both the position vector $\boldsymbol{r}$ and the velocity vector $\boldsymbol{v}$ thus lie in the plane perpendicular to $\boldsymbol{L}$. As $\boldsymbol{L}$ is conserved $\boldsymbol{r}$ and $\boldsymbol{v}$ must be confined to the plane perpendicular to $\boldsymbol{L}$ and through the origin. □

Applying the results of the previous section to the motion of a single particle under the action of a central force, we find (for the plane in which the particle moves):

$$F(r) = F_r = m\ddot{r} - mr\dot{\theta}^2 = m\ddot{r} - \frac{L^2}{mr^3}, \tag{6.9}$$

where we used that for a single particle, the magnitude of the angular momentum is given by $L = mr^2\dot{\theta}$. Rewriting equation (6.9) gives

$$m\ddot{r} = F(r) + \frac{L^2}{mr^3} = F(r) + F_{\mathrm{cf}}, \tag{6.10}$$

where F_{cf} is known as the *centrifugal force*, as it tends to move our particle away from the origin. We can write the centrifugal force as the derivative of a potential:

$$F_{\mathrm{cf}} = -\frac{\mathrm{d}U_{\mathrm{cf}}}{\mathrm{d}r} = -\frac{\mathrm{d}}{\mathrm{d}r}\left(\frac{L^2}{2mr^2}\right). \tag{6.11}$$

Writing the original force as the derivative of a potential $U(r)$ as well, we can write down an equation for the total energy of the system:

$$E = K + U = \frac{1}{2}m\dot{r}^2 + U(r) + \frac{L^2}{2mr^2}. \tag{6.12}$$

For both Newtonian gravity and the Coulomb force, the potential can be written as $U(r) = -\alpha/r$, where $\alpha = Gm_1m_2$ for gravity and $\alpha = -k_e q_1 q_2$ for Coulomb's law. We can then rewrite the energy equation as a differential equation for $r(t)$:

$$\frac{1}{2}m\left(\frac{\mathrm{d}r}{\mathrm{d}t}\right)^2 = E + \frac{\alpha}{r} - \frac{L^2}{2mr^2}. \tag{6.13}$$

To describe the motion of the particle, rather than specifying $r(t)$ and $\theta(t)$, we would like to express r as a function of θ. We can rewrite equation (6.13) to a differential equation for $r(\theta)$ by invoking the chain rule:

$$\left(\frac{\mathrm{d}r}{\mathrm{d}t}\right)^2 = \left(\frac{\mathrm{d}r}{\mathrm{d}\theta}\frac{\mathrm{d}\theta}{\mathrm{d}t}\right)^2 = \left(\frac{\mathrm{d}r}{\mathrm{d}\theta}\right)^2\left(\frac{L}{mr^2}\right)^2, \tag{6.14}$$

where we again used that $L = mr^2\dot{\theta}$. Equation (6.13) now becomes:

$$\left(\frac{1}{r^2}\frac{\mathrm{d}r}{\mathrm{d}\theta}\right)^2 = -\frac{1}{r^2} + \frac{2m\alpha}{L^2 r} + \frac{2mE}{L^2}. \tag{6.15}$$

Equation (6.15) isn't pretty, but can be simplified a little by absorbing the $1/r^2$ on the left hand side into the derivative, and completing the square on the right hand side:

$$\left(\frac{\mathrm{d}(\frac{1}{r})}{\mathrm{d}\theta}\right)^2 = -\left(\frac{1}{r} - \frac{m\alpha}{L^2}\right)^2 + \left(\frac{m\alpha}{L^2}\right)^2\left(1 + \frac{2EL^2}{m\alpha^2}\right). \tag{6.16}$$

We can simplify equation (6.16) further by introducing a new variable, $z = \frac{1}{r} - m\alpha/L^2$. We also introduce a dimensionless constant $\varepsilon = \sqrt{1 + 2EL^2/m\alpha^2}$ and an inverse length $q = m\alpha\varepsilon/L^2$. With these substitutions, our equation becomes:

$$\left(\frac{\mathrm{d}z}{\mathrm{d}\theta}\right)^2 = -z^2 + q^2. \tag{6.17}$$

We can solve equation (6.17) by separation of variables:

$$\int \frac{1}{\sqrt{q^2 - z^2}}\,\mathrm{d}z = \int \mathrm{d}\theta \Rightarrow \arccos\left(\frac{z}{q}\right) = \theta - \theta_0. \tag{6.18}$$

Taking the reference angle θ_0 (our integration constant) to be zero, we find $z(\theta) = q\cos(\theta)$. Translating back to $r(\theta)$, we obtain a fairly simple solution:

$$r(\theta) = \frac{L^2}{m\alpha}\frac{1}{1+\varepsilon\cos\theta}. \tag{6.19}$$

What the solution (6.19) (the *orbit* of our particle under the action of the central force) actually looks like, depends on the value of our dimensionless variable ε, known as the *eccentricity* of the orbit. To find out which orbits we can get, we translate equation (6.19) back to Cartesian coordinates, using $x = r\cos\theta$. Defining $k = L^2/m\alpha$, we get $k = r + \varepsilon r\cos\theta = r + \varepsilon x$, or $r = k - \varepsilon x$. Now using $r^2 = x^2 + y^2$, we get

$$x^2 + y^2 = (k - \varepsilon x)^2 = k^2 - 2\varepsilon kx + \varepsilon^2 x^2. \tag{6.20}$$

We can now distinguish four possibilities (see figure 6.2a):

1. $\varepsilon = 0$: In this case, equation (6.20) becomes $x^2 + y^2 = k^2$, so our orbit is a *circle* with the origin at its center.

2. $0 < \varepsilon < 1$: For this case, with some algebra (see problem 6.2), we can rewrite equation (6.20) as $((x - x_0)/a)^2 + (y/b)^2 = 1$, where $a = k/(1-\varepsilon^2)$, $x_0 = -\varepsilon a$, and $b = k/\sqrt{1-\varepsilon^2}$. These orbits are *ellipses*, with the center of the ellipse at $(x_0, 0)$, semi-major axis a, semi-minor axis b, and focal length $f = \sqrt{a^2 - b^2} = k\varepsilon/(1-\varepsilon^2) = -x_0$. One of the foci thus lies at the origin.

3. $\varepsilon = 1$: Equation (6.20) now becomes $y^2 = k^2 - 2kx$, which is the equation for a *parabola* (extending along the negative x-axis) with its 'top' (in this case, rightmost point) at $(k/2, 0)$ and focal length $k/2$, so the (single) focus is again located at the origin.

4. $\varepsilon > 1$: This case again requires some algebra to rewrite equation (6.20) in a recognizable standard form: $((x - x_0)/a)^2 - (y/b)^2 = 1$, where $a = k/(\varepsilon^2 - 1)$, $x_0 = \varepsilon a$ and $b = k/\sqrt{\varepsilon^2 - 1}$. These orbits are *hyperbola*, crossing the x-axis at $(x_0, 0)$, and approaching asymptotes $y = \pm b((x/a) - \varepsilon)$, which meet at $(x_0 + a, 0)$. The focal length is now $f = \sqrt{a^2 + b^2} = k\varepsilon/(\varepsilon^2 - 1) = \varepsilon a = x_0 + a$, so the focus of the hyperbola is also located at the origin.

In mathematics, these four possible types of orbits are known as *conic sections*: the curves you can get by intersecting a cone with a plane. Specifically, when the central force is gravity, such as in the solar system (where the sun is so much heavier than everything else combined that to good approximation we can describe orbits as being determined by the sun's gravitational field alone), the four cases listed above are the only possible orbits bodies can have. The planets, dwarf planets, asteroids, and many minor objects in our solar system all follow elliptical orbits around the sun, albeit with different eccentricity[1] - Earth's is almost zero (0.017), but that of Mars is significantly higher (0.09), and of Pluto much higher still (0.25). Comets, on the other hand, typically parabolic or hyperbolic orbits. Spacecraft such as the Voyager and New Horizons probes are often put on trajectories past planets that are not their final destination, to pick up (or loose) speed through a gravitational assist (in which they take a little bit of momentum from the planet's orbit); those paths past planets are typically hyperbola. Getting a spacecraft to orbit another planet (i.e., in a bound, so elliptical) orbit is actually much harder, but again, the resulting orbit is described by the maths presented above.

From the dimensionless eccentricity of an orbit, we can work out the total energy of that orbit easily. As $\varepsilon = \sqrt{1 + 2EL^2/m\alpha^2}$, we find

$$E_{\text{orbit}} = -\frac{1}{2}\left(\frac{\alpha}{L}\right)^2 (1-\varepsilon^2)m. \tag{6.21}$$

For elliptical orbits (where $(1-\varepsilon^2) = k/a = L^2/(\alpha m a)$, with a the semi-major axis), this expression gives a negative total energy that depends on the length of the semi-major axis only

$$E_{\text{ellipticalorbit}} = -\frac{\alpha}{2a}, \tag{6.22}$$

which is exactly half the potential energy of such an orbit. For a parabolic orbit (the boundary case between a closed and an open orbit), $\varepsilon = 1$, and so the total energy of the orbit vanishes; for a hyperbolic orbit we have $(\varepsilon^2 - 1) = k/a$ and we get a positive energy total energy.

[1] See table B.4 for data on the orbits of all planets and a number of their moons.

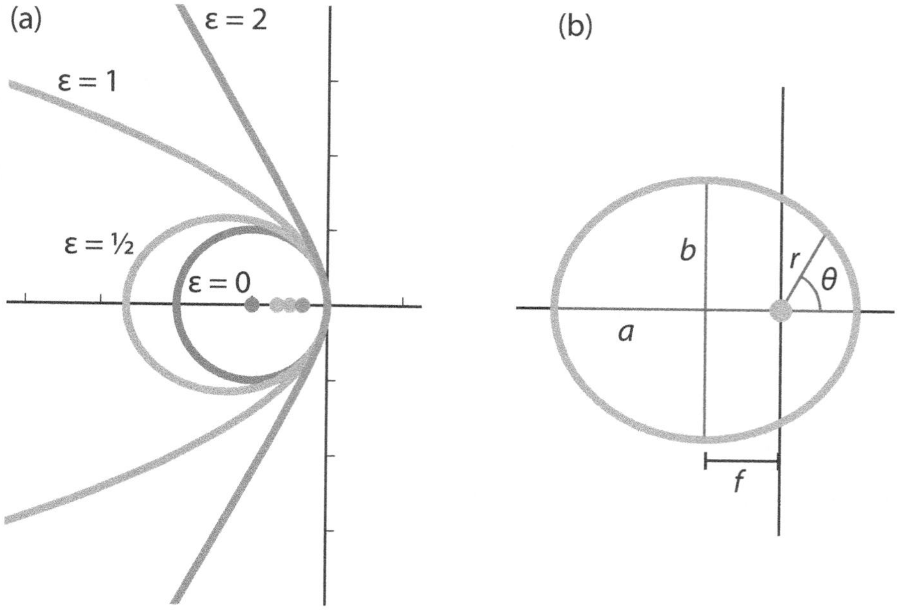

Figure 6.2: Orbits under the action of a central force. All orbits are described by equations (6.19) and (6.20) and are conic sections. (a) Four qualitatively different orbits, corresponding to different values of the parameter ε. Foci (colored dots) are shifted to show the orbits more clearly; as solutions of equation (6.19) or (6.20) all foci would be in the center. Blue: $\varepsilon = 0$, circular orbit. Orange: $\varepsilon = \frac{1}{2}$, elliptical orbit. Green: $\varepsilon = 1$, parabolic orbit. Red: $\varepsilon = 2$, hyperbolic orbit. (b) Properties of the elliptical orbit. The orbit has semi-major axis a, semi-minor axis b, and focal distance $f = \sqrt{a^2 - b^2}$. Distances in equation (6.19) are measured from the focal point in the origin; angles are measured with respect to the positive x-axis.

6.4 KEPLER'S LAWS

The fact that the planets move in elliptical orbits was first discovered by Kepler, based on observational data alone (he didn't have the benefit, as we do, of living after Newton and thus knowing about Newton's law of gravity). Kepler summarized his observational facts in three laws, which we can, with the benefit of hindsight, prove to be corollaries of Newton's laws.

Theorem 6.2 (Kepler's first law). *The planets move in elliptical orbits, with the sun at one of the foci.*

Proof. This is case two of the general result given by equations (6.19) and (6.20). □

Theorem 6.3 (Kepler's second law). *A line segment joining a planet and the sun sweeps out equal areas during equal intervals of time.*

Proof. This law is nothing but a special case of conservation of angular momentum. Consider a small piece of the orbit, in which the planet moves a distance $\mathrm{d}x$. The lines connecting the initial and final points of this piece of orbit with the sun make an angle $\mathrm{d}\theta$. If the initial distance from the planet to the sun was r, and the final distance $r + \mathrm{d}r$, we have, to first order, $\mathrm{d}x = r\,\mathrm{d}\theta$. The infinitesimal area the planet has swiped out is then given by (area of a triangle): $\mathrm{d}A = \frac{1}{2}r\,\mathrm{d}x = \frac{1}{2}r^2\,\mathrm{d}\theta$. If we want to know how much area was swept out over an amount of time, we need to know the time derivative of A, which is thus given by $\mathrm{d}A/\mathrm{d}t = \frac{1}{2}r^2\,\mathrm{d}\theta/\mathrm{d}t$. Now using that the angular momentum of the planet is given by $L = mr^2\dot{\theta}$, we find

$$\frac{\mathrm{d}A}{\mathrm{d}t} = \frac{r^2}{2}\frac{\mathrm{d}\theta}{\mathrm{d}t} = \frac{L}{2m}, \tag{6.23}$$

which is constant if L is conserved. □

Theorem 6.4 (Kepler's third law). *The square of the period T of an orbit is proportional to the cube of its semi-major-axis length a:*

$$T^2 = \frac{4\pi^2}{GM_\odot}a^3, \tag{6.24}$$

where $M_\odot$ is the mass of the sun.

Johannes Kepler (1571-1630) was a German astronomer and mathematician who made major contributions to understanding the motion of the planets. Copernicus had published his heliocentric (rather than geocentric) view of the universe posthumously in 1543, but the two systems were still heavily debated in Kepler's time. Having been convinced that Copernicus was right, Kepler set out to construct a geometric description of the solar system. He initially tried to do so using polyhedra and Platonic solids, but found that these could not accurately describe the data. In 1600, Kepler met with astronomer Tycho Brahe, who had made meticulous observations of the known planets, and, having been convinced of Kepler's skills in mathematics, shared his data with him. After Tycho's death in 1601, Kepler succeeded him as imperial mathematician in Prague, where he developed his laws over the next decade. Unfortunately, Kepler's Calvinist views got him in trouble frequently with both the Catholic and the Lutheran church, which led to his excummunication, but he managed to avoid further persecution by moving frequently, and he always could continue his scientific work. The Kepler spacecraft and mission, launched in 2009 to hunt for extrasolar terrestial planets, is named in his honor.

Figure 6.3: Portrait of Johannes Kepler (1610) [21].

Proof. We integrate equation (6.23) over the period of a whole orbit, which gives $A = LT/2m$. By Kepler's first law, the orbit is an ellipse, so its area equals $A = \pi ab$, with a and b the ellipse's semi-major and semi-minor axes. The two axes are related by $b = a\sqrt{1-\varepsilon^2}$, with ε again the eccentricity of the ellipse. Making these substitutions and squaring the resulting relation, we get:

$$\pi^2 a^4 = \frac{L^2}{m(1-\varepsilon^2)}\frac{T^2}{4m}. \tag{6.25}$$

Using $k = L^2/m\alpha$, like in equation (6.20), and the observation that for an elliptical orbit $k/(1-\varepsilon^2) = a$, we get $L^2/m(1-\varepsilon^2) = \alpha a$. Now for orbits in the solar system, $\alpha = GM_\odot m$, so we arrive at equation (6.24). □

Kepler's third law allows us to calculate the mass of any celestial body that is orbited by a (natural or artificial) satellite, simply by measuring the period and semi-major axis of the satellite's orbit.

6.5 PROBLEMS

6.1 A particularly useful orbit for satellites is the *geosynchronous* one: the orbit in which the satellite rotates around the Earth in exactly one day, so with respect to the ground, it is always in the same position. Find the altitude (i.e., distance above the Earth's surface) for a circular geosynchronous orbit.

6.2 In section 6.3, we found four possibilities for the orbits under a central force, all described by equation (6.19): $r(\theta) = k/(1+\varepsilon\cos\theta)$. For $0<\varepsilon<1$ these orbits are ellipses. In this problem, we'll find expressions for the quantities that we usually use to parametrize an ellipse (see figure 6.2b) in terms of ε and k.

(a) The easiest parameter to find is the length a of the semi-major axis, as twice the semi-major axis equals the distance between $r(0)$ and $r(\pi)$. From this equality, show that $a = k/(1-\varepsilon^2)$.

(b) The focal distance f is the distance from the center of the ellipse to the focal point. As we put the focal point in the origin, we have $f = a - r(0)$. Show that $f = \varepsilon a = k\varepsilon/(1-\varepsilon^2)$.

(c) Find x_0 (no extra calculation required).

(d) At the 'top' of the ellipse (where $y = b$), we have $\cos(\theta-\pi) = f/r$ and $r^2 = b^2 + f^2$. From these two equations, show that $b = a\sqrt{1-\varepsilon^2} = k/\sqrt{1-\varepsilon^2}$. *Hint: first show that at this position,* $r = a$*; this also gives that* $f = \sqrt{a^2-b^2}$.

6.3 Prove that for an elliptical orbit, the speed of the orbiting object is purely tangential at both the pericenter and apocenter (the closest and furthest point from the origin in the orbit). *Hint: take the time derivative of equation (6.19) to find the radial speed.*

6.4 At the time of writing, the Juno spacecraft is in a polar orbit around Jupiter. It's perijove (closest distance to Jupiter) is at an altitude of 4200 km, or 75600 km from Jupiter's center; Juno's apojove is $8.1\cdot10^6$ km from Jupiter's center. The spacecraft weighs 1593 kg. Jupiter's mass (known from the orbits of its moons) is $1.90\cdot10^{27}$ kg.

(a) Find the semi-major axis and eccentricity of Juno's orbit around Jupiter.

(b) How long does it take Juno to complete one orbit?

(c) Find Juno's maximum and minimum speed while traversing this orbit. You may find the result of problem 6.3 useful in answering this question.

(d) Suppose NASA wants to make Juno's orbit more circular. Should it have Juno fire its thruster rockets to accelerate or decelerate its angular velocity?

6.5 Kepler's laws apply to the case that an object with relatively small mass m orbits an object with large mass M, which we assume stays fixed. Technically, this is incorrect: both objects rotate about their common center of mass. Fortunately, we can still use the expressions derived in this section, with a small modification. To see how this works, we write down the equations of motion for the two objects, due to the force they exert on each other:

$$\ddot{\boldsymbol{x}}_1 = -\frac{1}{m}\boldsymbol{F}(\boldsymbol{r}), \quad \ddot{\boldsymbol{x}}_2 = \frac{1}{M}\boldsymbol{F}(\boldsymbol{r}), \tag{6.26}$$

where $\boldsymbol{x}_1$ is the position of the object with mass m, $\boldsymbol{x}_2$ that of the object with mass M, and $\boldsymbol{r} = \boldsymbol{x}_2 - \boldsymbol{x}_1$ their separation. We denote the position of the center of mass of the system by $\boldsymbol{R}$.

(a) As there is no external force acting on the system, the total momentum is conserved and therefore the center of mass cannot accelerate. Argue that this implies that

$$(m+M)\ddot{\boldsymbol{R}} = 0, \tag{6.27}$$

and combine equations (6.26) and (6.27) into an expression for $\boldsymbol{R}$ in terms of $\boldsymbol{x}_1$, $\boldsymbol{x}_2$, and the masses of the two objects.

(b) From equation (6.26), also derive an equation of motion for the separation $\boldsymbol{r}$ between the two objects.

The equations you found in (a) and (b) together are equivalent to the equations of motion in (6.26), but only one is a differential equation, and they are uncoupled: we don't need to know the position of the center of mass to find the separation, and vice versa.

(c) Show that you can re-write the equation of motion for the separation between the two objects as $\boldsymbol{F}(\boldsymbol{r}) = \mu\ddot{\boldsymbol{r}}$, where μ is the *reduced mass* that we also encountered when studying collisions in the center of mass frame, equation (4.36), given by

$$\mu = \frac{mM}{m+M}.$$

Note that solving the final equation for the separation $\boldsymbol{r}$ is entirely equivalent to solving the equation of motion of a single particle under the action of a central force, with the modification that the mass of the particle is replaced by the reduced mass. For the case that $m << M$, the reduced mass is approximately equal to m.

(d) Calculate the reduced mass of the Earth-Moon two body problem. Can we state that the Moon revolves around the Earth?

(e) Nowhere in the derivations in this problem did we assume that $m << M$. The same rules apply to any two objects. Consider the opposite limit: two objects (these might for instance be binary stars) of equal mass M that rotate around their common center of mass. Show that for this case, for circular orbits the orbital period is given by

$$T^2 = \frac{2\pi^2 d^3}{GM}, \tag{6.28}$$

where d is the distance between the two objects.

6.6 A student with mass 65.0 kg stands at the center of a simple merry-go-round that consists of a large disk of radius 1.5 m and mass 25 kg and is making a full rotation every 2.0 s. The student walks out to a distance of 0.50 m from the center.

(a) Find the rotational frequency of the merry-go-round with the student at this point.

(b) What are the forces acting on the student at this point?

7

GENERAL ROTATIONAL MOTION*

7.1 LINEAR AND ANGULAR VELOCITY

We related the linear and angular velocities of a rotating object in two dimensions in section 5.1. There, we also already stated the relation between the linear velocity vector and rotation vector in three dimensions (equation (5.5)):

$$\boldsymbol{v} = \boldsymbol{\omega} \times \boldsymbol{r}. \tag{7.1}$$

It is not hard to see that this expression indeed simplifies to the scalar relationship $v = \omega r$ for rotations in a plane, with the right sign for the linear velocity. That's hardly a proof though, so let's put this on some more solid footing. Suppose $\boldsymbol{r}$ makes an angle ϕ with $\boldsymbol{\omega}$. Suppose also that it changes by $\mathrm{d}\boldsymbol{r}$ in a time interval $\mathrm{d}t$, then if we have pure rotation, $\mathrm{d}\boldsymbol{r}$ is perpendicular to both $\boldsymbol{r}$ and $\boldsymbol{\omega}$, and its magnitude is given by $|\mathrm{d}\boldsymbol{r}| = \omega r \sin\phi \,\mathrm{d}t = |\boldsymbol{\omega} \times \boldsymbol{r}|\,\mathrm{d}t$, where ω and r are the lengths of their respective vectors. Finally, as seen from the top (i.e., looking down the vector $\boldsymbol{\omega}$), the rotation should be counter-clockwise (by definition of the direction of $\boldsymbol{\omega}$), which corresponds with the direction of $\boldsymbol{\omega} \times \boldsymbol{r}$. We thus find that both the magnitude and direction of $\boldsymbol{v} = \mathrm{d}\boldsymbol{r}/\mathrm{d}t$ indeed equal $\boldsymbol{\omega} \times \boldsymbol{r}$, and equation (5.5) holds.

7.2 ROTATING REFERENCE FRAMES

In section 4.3, we considered what happens if we considered the (linear) motion of an object from a stationary ('lab frame') or co-moving point of view, with special attention for the center of mass frame. These frames were moving with constant velocity with respect to each other, and were all inertial frames - Newton's first and second laws hold in all inertial frames. In this section, we'll consider a rotating reference frame, where instead of co-moving with a linear velocity, we co-rotate with a constant angular velocity. Rotating reference frames are not inertial frames, as to keep something rotating (and thus change the direction of the linear velocity) requires the application of a net force. Instead, as we'll see, in a rotating frame of reference we'll get all sorts of *fictitious forces* - forces that have no real physical source, like gravity or electrostatics, but originate from the fact that we're in a rotating reference frame.

In a rotating reference frame, the direction of the basis vectors changes over time (as measured with respect to the stationary lab frame - this is different from the linearly co-moving frames, where the directions of the basis vectors remained constant). To see how the basis vectors change over time, we can simply calculate their linear velocity, using equation (5.5):

$$\frac{\mathrm{d}\hat{\boldsymbol{x}}}{\mathrm{d}t} = \boldsymbol{\omega} \times \hat{\boldsymbol{x}}, \qquad \frac{\mathrm{d}\hat{\boldsymbol{y}}}{\mathrm{d}t} = \boldsymbol{\omega} \times \hat{\boldsymbol{y}}, \qquad \frac{\mathrm{d}\hat{\boldsymbol{z}}}{\mathrm{d}t} = \boldsymbol{\omega} \times \hat{\boldsymbol{z}}. \tag{7.2}$$

We can now easily determine the change of an arbitrary vector $\boldsymbol{u} = u_x\hat{\boldsymbol{x}} + u_y\hat{\boldsymbol{y}} + u_z\hat{\boldsymbol{z}}$ over time:

$$\begin{aligned} \frac{\mathrm{d}\boldsymbol{u}}{\mathrm{d}t} &= \left(\frac{\partial u_x}{\partial t}\hat{\boldsymbol{x}} + \frac{\partial u_y}{\partial t}\hat{\boldsymbol{y}} + \frac{\partial u_z}{\partial t}\hat{\boldsymbol{z}}\right) + \left(u_x\frac{\mathrm{d}\hat{\boldsymbol{x}}}{\mathrm{d}t} + u_y\frac{\mathrm{d}\hat{\boldsymbol{y}}}{\mathrm{d}t} + u_z\frac{\mathrm{d}\hat{\boldsymbol{z}}}{\mathrm{d}t}\right) \\ &= \frac{\delta \boldsymbol{u}}{\delta t} + \boldsymbol{\omega} \times \boldsymbol{u}, \end{aligned} \tag{7.3}$$

where $\frac{\delta \boldsymbol{u}}{\delta t}$ is defined by equation (7.3), and represents the time derivative of $\boldsymbol{u}$ in the rotating basis[1]. We see that in addition to the 'regular' time derivative, acting on the components of $\boldsymbol{u}$, we get an additional term $\boldsymbol{\omega} \times \boldsymbol{u}$ due to the rotation of the system. Note that for $\boldsymbol{u} = \boldsymbol{\omega}$, we find that $\mathrm{d}\boldsymbol{\omega}/\mathrm{d}t = \delta\boldsymbol{\omega}/\delta t = \dot{\boldsymbol{\omega}}$, i.e., the time derivative of the rotation vector is the same in the stationary and rotating frames.

The prime example of a vector is of course the position vector $\boldsymbol{r}$ of a particle, the second derivative of which appears in Newton's second law of motion. We'll calculate that second derivative for a position vector in a rotating coordinate frame. The first derivative is a simple application of equation (7.3):

$$\frac{\mathrm{d}\boldsymbol{r}}{\mathrm{d}t} = \frac{\delta \boldsymbol{r}}{\delta t} + \boldsymbol{\omega} \times \boldsymbol{r}. \tag{7.4}$$

To get the second derivative, we apply (7.3) to the velocity vector found in (7.4):

$$\begin{aligned} \frac{\mathrm{d}^2\boldsymbol{r}}{\mathrm{d}t^2} &= \frac{\mathrm{d}}{\mathrm{d}t}\left(\frac{\delta \boldsymbol{r}}{\delta t} + \boldsymbol{\omega} \times \boldsymbol{r}\right) \\ &= \frac{\delta}{\delta t}\left(\frac{\delta \boldsymbol{r}}{\delta t} + \boldsymbol{\omega} \times \boldsymbol{r}\right) + \boldsymbol{\omega} \times \left(\frac{\delta \boldsymbol{r}}{\delta t} + \boldsymbol{\omega} \times \boldsymbol{r}\right) \\ &= \frac{\delta^2 \boldsymbol{r}}{\delta t^2} + \frac{\delta \boldsymbol{\omega}}{\delta t} \times \boldsymbol{r} + 2\boldsymbol{\omega} \times \frac{\delta \boldsymbol{r}}{\delta t} + \boldsymbol{\omega} \times (\boldsymbol{\omega} \times \boldsymbol{r}). \end{aligned} \tag{7.5}$$

Like in the two-dimensional case given by equation (6.4), we find that the acceleration in a rotating reference frame picks up extra terms compared to a stationary (or more general, inertial) frame. To get a complete picture, we also allow the origin of the rotating frame to be different from that of the stationary lab frame. Let $\boldsymbol{r}_{\mathrm{lab}}$ be the position vector in the lab frame, $\boldsymbol{R}$ the vector pointing from the origin of the lab frame to that of the rotating frame, and $\boldsymbol{r}$ the position vector in the rotating frame. We then have $\boldsymbol{r}_{\mathrm{lab}} = \boldsymbol{R} + \boldsymbol{r}$, and for the second derivative of $\boldsymbol{r}_{\mathrm{lab}}$ we find:

$$\begin{aligned} \frac{\mathrm{d}^2\boldsymbol{r}_{\mathrm{lab}}}{\mathrm{d}t^2} &= \frac{\mathrm{d}^2\boldsymbol{r}}{\mathrm{d}t^2} + \frac{\mathrm{d}^2\boldsymbol{R}}{\mathrm{d}t^2} \\ &= \frac{\delta^2 \boldsymbol{r}}{\delta t^2} + \boldsymbol{\omega} \times (\boldsymbol{\omega} \times \boldsymbol{r}) + 2\boldsymbol{\omega} \times \frac{\delta \boldsymbol{r}}{\delta t} + \frac{\delta \boldsymbol{\omega}}{\delta t} \times \boldsymbol{r} + \frac{\mathrm{d}^2\boldsymbol{R}}{\mathrm{d}t^2}. \end{aligned} \tag{7.6}$$

We can substitute equation (7.6) in Newton's second law of motion in the lab frame (i.e., just $\boldsymbol{F} = \mathrm{d}\boldsymbol{r}_{\mathrm{lab}}/\mathrm{d}t$) to find the expression for that law in the rotating frame:

$$m\frac{\delta^2 \boldsymbol{r}}{\delta t^2} = \boldsymbol{F} - m\left[\boldsymbol{\omega} \times (\boldsymbol{\omega} \times \boldsymbol{r}) + 2\boldsymbol{\omega} \times \boldsymbol{v} + \dot{\boldsymbol{\omega}} \times \boldsymbol{r} + \frac{\mathrm{d}^2\boldsymbol{R}}{\mathrm{d}t^2}\right], \tag{7.7}$$

where we defined $\boldsymbol{v} = \delta\boldsymbol{r}/\delta t$ as the velocity in the rotating frame, and used that the time derivative of $\boldsymbol{\omega}$ is the same in both the stationary and the rotating frame. We find that we get four correction terms to the force due to our transition to a rotating frame. They are not 'real' forces like gravity or friction, as they vanish in the lab frame, but you can easily experience their effects, when you're in a turning car or rotating carousel. As they have no physical origin, we call these forces *fictitious*. They are known as the centrifugal, Coriolis, azimuthal, and translational force, respectively:

$$\boldsymbol{F}_{\mathrm{cf}} = -m\boldsymbol{\omega} \times (\boldsymbol{\omega} \times \boldsymbol{r}) \tag{7.8}$$

$$\boldsymbol{F}_{\mathrm{Cor}} = -2m\boldsymbol{\omega} \times \boldsymbol{v} \tag{7.9}$$

$$\boldsymbol{F}_{\mathrm{az}} = -m\dot{\boldsymbol{\omega}} \times \boldsymbol{r} \tag{7.10}$$

$$\boldsymbol{F}_{\mathrm{trans}} = -m\frac{\mathrm{d}^2\boldsymbol{R}}{\mathrm{d}t^2} \tag{7.11}$$

We encountered the centrifugal, Coriolis and azimuthal force before in section 6.2. To see how the expressions above connect to the planar versions, let us pick the coordinates of the rotating frame such that the direction of $\boldsymbol{\omega}$ coincides with the z-axis. The rotational motion and the forces can then be described in terms of the cylindrical coordinates consisting of the polar coordinates (ρ, θ) in the xy-plane and z along the z-axis (note that we use $\rho = \sqrt{x^2 + y^2}$ for the radial distance in the xy plane instead of r, as r is now the distance

[1]Some authors use the notation $\left(\frac{\mathrm{d}\boldsymbol{u}}{\mathrm{d}t}\right)_{\mathrm{rot}}$ for $\frac{\delta \boldsymbol{u}}{\delta t}$.

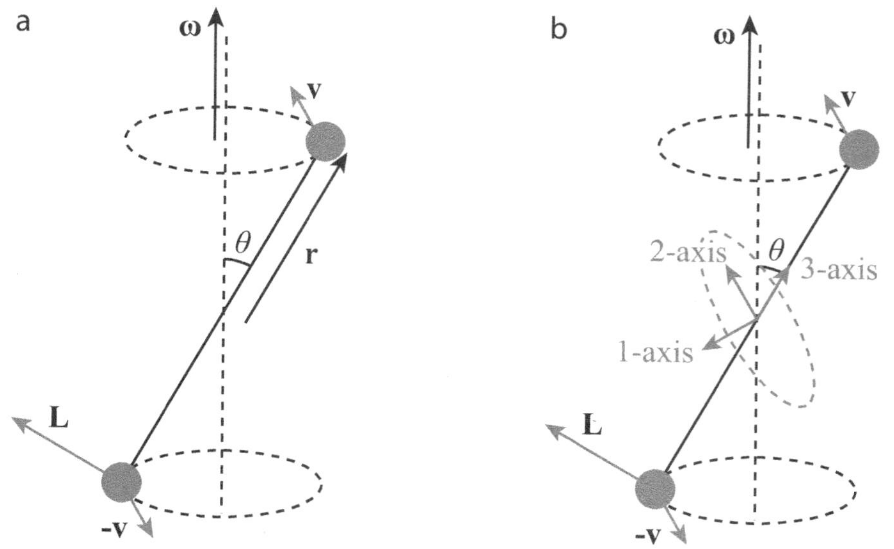

Figure 7.1: Rotation of a dumbbell, consisting of two equal masses m separated by a distance $d = 2r$ about an axis which is not a symmetry axis of the system.

from the origin to our point in three dimensions). For the centrifugal force we have $\boldsymbol{\omega} \cdot \boldsymbol{F}_{\text{cf}} = 0$, so it lies in our newly defined xy plane, and in cylindrical coordinates it can be expressed as

$$\boldsymbol{F}_{\text{cf}} = -m\left[\boldsymbol{\omega}(\boldsymbol{\omega} \cdot \boldsymbol{r}) - \omega^2 \boldsymbol{r}\right] = m\omega^2(x\hat{\boldsymbol{x}} + y\hat{\boldsymbol{y}}) = m\omega^2 \rho\hat{\boldsymbol{\rho}}. \tag{7.12}$$

The centrifugal force is thus nothing but minus the centripetal force, which we already encountered for uniform rotational motion in equation (5.10), and as the second term in (6.6). The centrifugal force is the force you 'feel' pushing you sidewards when your car makes a sharp turn, and is also responsible for creating the parabolic shape of the water surface in a spinning bucket, see problem 7.4.

The Coriolis force is present whenever a particle is moving with respect to the rotating coordinates, and tends to deflect particles from a straight line (which you'd get in an inertial reference frame). We have $\boldsymbol{\omega} \cdot \boldsymbol{F}_{\text{Cor}} = \boldsymbol{v} \cdot \boldsymbol{F}_{\text{Cor}} = 0$, so the Coriolis force is perpendicular to both the rotation and velocity vectors - note that this is the velocity in the rotating frame. In the two-dimensional case, we had $\boldsymbol{F}_{\text{Cor}} = 2m\dot{\rho}\omega\hat{\boldsymbol{\theta}}$ (equation 6.8), which for a velocity in the radial direction, $\dot{\rho}$, gives a force in the angular direction $\hat{\boldsymbol{\theta}}$.

The azimuthal force occurs when the rotation vector of our rotating system changes - i.e., when the rotation speeds up or decelerates, or the plane of rotation alters. In either case we have $\boldsymbol{r} \cdot \boldsymbol{F}_{\text{Cor}} = 0$, so the force is perpendicular to the position vector. If it is the magnitude of the rotation vector that changes, and we again take the rotation vector to lie along the z-axis in the rotating frame, $\dot{\boldsymbol{\omega}}$ also lies along the z-axis, and we get

$$\boldsymbol{F}_{\text{az}} = -m\dot{\omega}\hat{\boldsymbol{z}} \times \boldsymbol{r} = -mr\ddot{\theta}\hat{\boldsymbol{\theta}}, \tag{7.13}$$

so the azimuthal force is minus m times the tangential acceleration α (see equation 5.8), or minus the first term of (6.7).

The translational force finally occurs when the rotating reference frame's origin accelerates with respect to that of the stationary lab frame. You also feel it if the 'rotating' reference frame is actually not rotating, but only accelerating linearly - it's the force that pushes you back in your seat when your car or train accelerates.

7.3 Rotations about an arbitrary axis

7.3.1 Moment of inertia tensor

In chapter 5, we studied the rotation of rigid bodies about an axis of symmetry. For these cases, we have $\boldsymbol{L} = I\boldsymbol{\omega}$, where I is the moment of inertia with respect to the rotation axis. We already noted that I depends on which axis we pick, and that the proportional relation between the rotation vector and angular momentum is not the most general possibility. In this section, we'll derive the more general form, in which the number I is replaced by a 2-tensor, i.e., a map from a vector space (here $\mathbb{R}^3$) into itself, represented by a 3×3 matrix.

To arrive at the more general relation between $\boldsymbol{L}$ and $\boldsymbol{\omega}$, we go back to the original definition of $\boldsymbol{L} = \boldsymbol{r} \times \boldsymbol{p}$, and consider the motion of a dumbbell around an axis which is not a symmetry axis (see figure 7.1). If the dumbbell makes an angle θ with the rotation axis, and rotates counter-clockwise as seen from the top, we get:

$$\begin{aligned} \boldsymbol{L} &= m\boldsymbol{r} \times \boldsymbol{v} + m(-\boldsymbol{r}) \times (-\boldsymbol{v}) \\ &= 2m\boldsymbol{r} \times (\boldsymbol{\omega} \times \boldsymbol{r}) \\ &= 2m\left[\boldsymbol{\omega}(\boldsymbol{r}\cdot\boldsymbol{r}) - \boldsymbol{r}(\boldsymbol{r}\cdot\boldsymbol{\omega})\right] \\ &= 2mr^2\boldsymbol{\omega} - 2m\omega r\cos\theta\,\boldsymbol{r} \end{aligned} \tag{7.14}$$

where we used equation (5.5) relating the linear velocity $\boldsymbol{v}$ to the rotational velocity $\boldsymbol{\omega}$ through $\boldsymbol{v} = \boldsymbol{\omega} \times \boldsymbol{r}$. Equation (7.14) shows that for a rotation about an arbitrary axis through the center of the dumbbell, we get two terms in $\boldsymbol{L}$. The first term, $2mr^2\boldsymbol{\omega}$, is the rotation about an axis perpendicular to the dumbbell, and equals $I\boldsymbol{\omega}$ for $I = 2mr^2$, as we found in section 5.4. The second term, $-2m\omega r\cos\theta\boldsymbol{r}$, tells us that in general we also get a component of $\boldsymbol{L}$ along the axis pointing from the rotation center to the rotating mass (i.e., the arm). Note that the two terms cancel when $\theta = \pi/2$, as we'd expect for in that case the moment of inertia of the dumbbell is zero.

We can easily generalize equation (7.14) to any set of masses m_α with position vectors $\boldsymbol{r}_\alpha$ (where the index α runs over all particles), and with a rotation $\boldsymbol{\omega}$ about an arbitrary axis:

$$\boldsymbol{L} = \sum_\alpha m_\alpha \boldsymbol{r}_\alpha \times (\boldsymbol{\omega} \times \boldsymbol{r}_\alpha) \equiv \boldsymbol{I}\cdot\boldsymbol{\omega}. \tag{7.15}$$

The *moment of inertia tensor* is defined by equation (7.15). It is a symmetric tensor, mapping a vector $\boldsymbol{\omega}$ in $\mathbb{R}^3$ onto another vector $\boldsymbol{L}$ in $\mathbb{R}^3$. In Cartesian coordinates, we can express its nine components as three moments of inertia about the x, y and z axes, which will be the diagonal terms of $\boldsymbol{I}$:

$$I_{xx} = \sum_\alpha m_\alpha(y_\alpha^2 + z_\alpha^2) \qquad I_{yy} = \sum_\alpha m_\alpha(x_\alpha^2 + z_\alpha^2) \qquad I_{zz} = \sum_\alpha m_\alpha(x_\alpha^2 + y_\alpha^2), \tag{7.16}$$

and three *products of inertia* for the off-diagonal components:

$$I_{xy} = I_{yx} = -\sum_\alpha m_\alpha x_\alpha y_\alpha \qquad I_{xz} = I_{zx} = -\sum_\alpha m_\alpha x_\alpha z_\alpha \qquad I_{yz} = I_{zy} = -\sum_\alpha m_\alpha y_\alpha z_\alpha. \tag{7.17}$$

We can also write equations (7.16) and (7.17) more succinctly using index notation, where i and j run over x, y and z, and we use the Kronecker delta δ_{ij} which is one if $i = j$ and zero if $i \neq j$:

$$I_{ij} = \sum_\alpha m_\alpha(r^2\delta_{ij} - r_i r_j). \tag{7.18}$$

Equations (7.16) and (7.17) generalize to continuous objects in the same way equation (5.13) generalized to (5.14). Using the index notation again, we can explicitly write:

$$I_{ij} = \int_V (r^2\delta_{ij} - r_i r_j)\rho(\boldsymbol{r})\,\mathrm{d}V. \tag{7.19}$$

The moment of inertia tensor contains all information about the rotational inertia of an object (or a collection of particles) about any axis. In particular, if one of the axes (say the z-axis) is an axis of symmetry, we get that $I_{xz} = I_{yz} = 0$, and for rotations about that axis (so $\boldsymbol{\omega} = \omega\hat{\boldsymbol{z}}$), we retrieve $\boldsymbol{L} = I_z\boldsymbol{\omega}$.

In addition to calculating the angular momentum, we can also use the moment of inertia tensor for calculating the kinetic energy for rotations about an arbitrary axis. We have:

$$\begin{aligned} K &= \frac{1}{2}\sum_\alpha m_\alpha \boldsymbol{v}_\alpha \cdot \boldsymbol{v}_\alpha \\ &= \frac{1}{2}\sum_\alpha m_\alpha(\boldsymbol{\omega}\times\boldsymbol{r}_\alpha)\cdot(\boldsymbol{\omega}\times\boldsymbol{r}_\alpha) \\ &= \frac{1}{2}\boldsymbol{\omega}\cdot\left[\sum_\alpha m_\alpha\boldsymbol{r}_\alpha\times(\boldsymbol{\omega}\times\boldsymbol{r}_\alpha)\right] \\ &= \frac{1}{2}\boldsymbol{\omega}\cdot\boldsymbol{L} = \frac{1}{2}\boldsymbol{\omega}\cdot\boldsymbol{I}\cdot\boldsymbol{\omega}. \end{aligned} \tag{7.20}$$

Leonhard Euler (1707-1783) was a Swiss mathematician who made major contributions to many different branches of mathematics, and, by application, physics. He also introduced much of the modern mathematical terminology and notation, including the concept of (mathematical) functions. Euler was possibly the most prolific mathematician who ever lived, and likely is the person with the most equations and formulas named after him. Although his father, who was a pastor, encouraged Euler to follow in his footsteps, Euler's tutor, famous mathematician (and family friend) Johann Bernoulli convinced both father and son that Euler's talent for mathematics would make him a giant in the field. Famous examples of Euler's work include his contributions to graph theory (the Köningsberg bridges problem), the relation $e^{i\pi}+1=0$ between five fundamental mathematical numbers named after him, his work on power series, a method for numerically solving differential equations, and his work on fluid mechanics (in which there is also an 'Euler's equation').

Figure 7.2: Portrait of Leonhard Euler by Jakob Handmann (1753) [22].

7.3.2 EULER'S EQUATIONS

In the lab frame, we have equation (5.25) relating the torque and the angular momentum. We used this equation to prove conservation of angular momentum in the absence of a net external torque, and to study precession. However, for rotations about an arbitrary axis, it is easier to transform to a frame in which we rotate with the object, much like moving with the center of mass makes the study of collisions much easier. We've already done the math for transforming to a co-rotating frame in section 7.2; here we only need the result in equation (7.3) to find the time derivative of the angular momentum in the rotating frame. Equation (5.25) then translates to:

$$\boldsymbol{\tau} = \frac{\delta \boldsymbol{L}}{\delta t} + \boldsymbol{\omega} \times \boldsymbol{L} = \boldsymbol{I} \cdot \dot{\boldsymbol{\omega}} + \boldsymbol{\omega} \times (\boldsymbol{I} \cdot \boldsymbol{\omega}). \tag{7.21}$$

Now since $\boldsymbol{I}$ is symmetric, all its eigenvalues are real, and its eigenvectors are a basis for $\mathbb{R}^3$; moreover, for distinct eigenvalues the eigenvectors are orthogonal, so from the eigenvector basis we can easily construct an orthonormal basis $(\hat{\boldsymbol{e}}_1, \hat{\boldsymbol{e}}_2, \hat{\boldsymbol{e}}_3)$ of eigenvectors corresponding to the three eigenvalues I_1, I_2 and I_3. If we express the moment of inertia tensor in this orthonormal eigenvector basis, its representation becomes a simple diagonal matrix, $\boldsymbol{I} = \mathrm{diag}(I_1, I_2, I_3)$. We call the directions $\hat{\boldsymbol{e}}_i$ the *principal axes* of our rotating object, and the associated eigenvalues the *principal moments of inertia*. The construction of the principal axes and moments of inertia works for any object - including ones that do not exhibit any kind of symmetry. If an object does have a symmetry axis, that axis is usually also a principal axis, as can easily be checked by calculating the products of inertia with respect to that axis (they vanish for a principal axis).

If we express our rotational quantities in the principal axis basis $\{\hat{\boldsymbol{e}}_i\}$ of our rotating object, our equations become much simpler. We have

$$\boldsymbol{L} = \boldsymbol{I} \cdot \boldsymbol{\omega} = \begin{pmatrix} I_1 & 0 & 0 \\ 0 & I_2 & 0 \\ 0 & 0 & I_3 \end{pmatrix} \begin{pmatrix} \omega_1 \\ \omega_2 \\ \omega_3 \end{pmatrix} = \begin{pmatrix} I_1\omega_1 \\ I_2\omega_2 \\ I_3\omega_3 \end{pmatrix}, \tag{7.22}$$

or in components: $L_i = I_i \omega_i$. Equation (7.21) simplifies to:

$$\boldsymbol{\tau} = \begin{pmatrix} I_1\dot{\omega}_1 \\ I_2\dot{\omega}_2 \\ I_3\dot{\omega}_3 \end{pmatrix} + \begin{pmatrix} \omega_1 \\ \omega_2 \\ \omega_3 \end{pmatrix} \times \begin{pmatrix} I_1\omega_1 \\ I_2\omega_2 \\ I_3\omega_3 \end{pmatrix}, \tag{7.23}$$

which gives for the three components of the torque:

$$\tau_1 = I_1\dot{\omega}_1 + (I_3 - I_2)\omega_3\omega_2, \tag{7.24a}$$
$$\tau_2 = I_2\dot{\omega}_2 + (I_1 - I_3)\omega_1\omega_3, \tag{7.24b}$$
$$\tau_3 = I_3\dot{\omega}_3 + (I_2 - I_1)\omega_2\omega_1. \tag{7.24c}$$

Equations (7.24) are known as *Euler's equations* (of a rotating object - the classification is necessary as there are many equations associated with Euler).

As an example, let's apply Euler's equations to our dumbbell. We take the origin at the pivot, i.e., where the rotation axis crosses the dumbbell's own axis. The dumbbell does have rotational symmetry, about the axis connecting the two masses - let's call that the 3-axis. The other two axes then span the plane perpendicular to the dumbbell; we can pick any orthonormal pair for the 1 and 2-axes. The rotation vector in this basis is given by

$$\boldsymbol{\omega} = \begin{pmatrix} \omega_1 \\ \omega_2 \\ \omega_3 \end{pmatrix} = \omega \begin{pmatrix} 0 \\ \sin\theta \\ \cos\theta \end{pmatrix}. \tag{7.25}$$

The products of inertia vanish; the principal moments are given by $I_1 = I_2 = \frac{1}{2}md^2$ (with d the distance between the two masses) and $I_3 = 0$. As long as the rotational velocity is constant ($\dot{\boldsymbol{\omega}} = 0$), we get from Euler's equations that $\tau_2 = \tau_3 = 0$, and $\tau_1 = -\frac{1}{2}md^2\omega^2 \sin\theta\cos\theta$. We can thus rotate our dumbbell about an axis that's not a symmetry axis, but at a price: it exerts a torque on its support, which in turn exerts a counter-torque to keep the dumbbell's rotation axis in place. This torque will change the angular momentum of our dumbbell over time. If we remove the force exerting the counter-torque (e.g., if our dumbbell is supported at the pivot, remove the support), the dumbbell will turn, in our example about axis 2, until the rotation vector $\boldsymbol{\omega}$ has become parallel with the angular momentum vector $\boldsymbol{L}$.

For the dumbbell, as it has a rotational symmetry, two of the principal moments are identical. There are many objects that have no such symmetry, but they still have three well-defined principal axes. While for the dumbbell rotation about any of the principal axes is stable, this is not the case for an object with three different principal moments. A good example is a tennis racket, whose principal axes are sketched in figure 7.3. The accompanying theorem about the stability of rotations about these axes is easily demonstrated with a tennis racket, and bears its name.

Theorem 7.1 (Tennis racket theorem). *If the three principal moments of inertia of an object are different (say $I_1 < I_2 < I_3$), then rotations about the principal axes 1 and 3 associated with the maximum and minimum moments I_1 and I_3 are stable, but those about the principal axis 2 associated with the intermediate moment I_2 are unstable.*

Proof. For rotations about a principal axis, the torque is zero (by construction), so Euler's equations read

$$\dot{\omega}_1 + \frac{I_3 - I_2}{I_1}\omega_3\omega_2 = 0, \tag{7.26a}$$

$$\dot{\omega}_2 + \frac{I_1 - I_3}{I_2}\omega_1\omega_3 = 0, \tag{7.26b}$$

$$\dot{\omega}_3 + \frac{I_2 - I_1}{I_3}\omega_2\omega_1 = 0. \tag{7.26c}$$

If we rotate about axis 1, then ω_2 and ω_3 are (at least initially) very small, so equation (7.26a) gives $\dot{\omega}_1 = 0$. We can then derive an equation for ω_2 by taking the time derivative of (7.26b) and using (7.26c) for $\dot{\omega}_3$, which gives:

$$0 = \ddot{\omega}_2 + \frac{I_1 - I_3}{I_2}(\dot{\omega}_1\omega_3 + \omega_1\dot{\omega}_3) = \ddot{\omega}_2 - \frac{I_1 - I_3}{I_2}\frac{I_2 - I_1}{I_3}\omega_1^2\omega_2. \tag{7.27}$$

Now $(I_1 - I_3)/I_2 < 0$, $(I_2 - I_1)/I_3 > 0$, and $\omega_1^2 > 0$, and ω_2 satisfies the differential equation $\ddot{\omega}_2 = -c\omega_2$, with $c > 0$. Solutions to this equation are of course sines and cosines with constant amplitude. Although ω_2 can thus be finite, its amplitude does not grow over time (and will in fact decrease due to drag), so rotations about axis 2 are opposed. Similarly, we find that rotations about axis 3 cannot grow in amplitude either, and rotations about axis 1 are stable. We can repeat the same argument for axes 2 and 3. For axis 3, we find that rotations about the other two axes are likewise opposed, so rotations about this axis are stable as well. For axis 2 on the other hand, we find that

$$0 = \ddot{\omega}_1 - \frac{I_3 - I_2}{I_1}\frac{I_2 - I_1}{I_3}\omega_2^2\omega_1, \tag{7.28}$$

or $\ddot{\omega}_1 = c\omega_1$, with c another positive constant. Solutions to this equation are not sines and cosines, but exponential: $\omega_1(t) = A\exp(\sqrt{c}t) + B\exp(-\sqrt{c}t)$, which means that for any finite initial rotation about the 1 axis, the amplitude of this rotation will grow over time, and rotations about the 2-axis are thus unstable. □

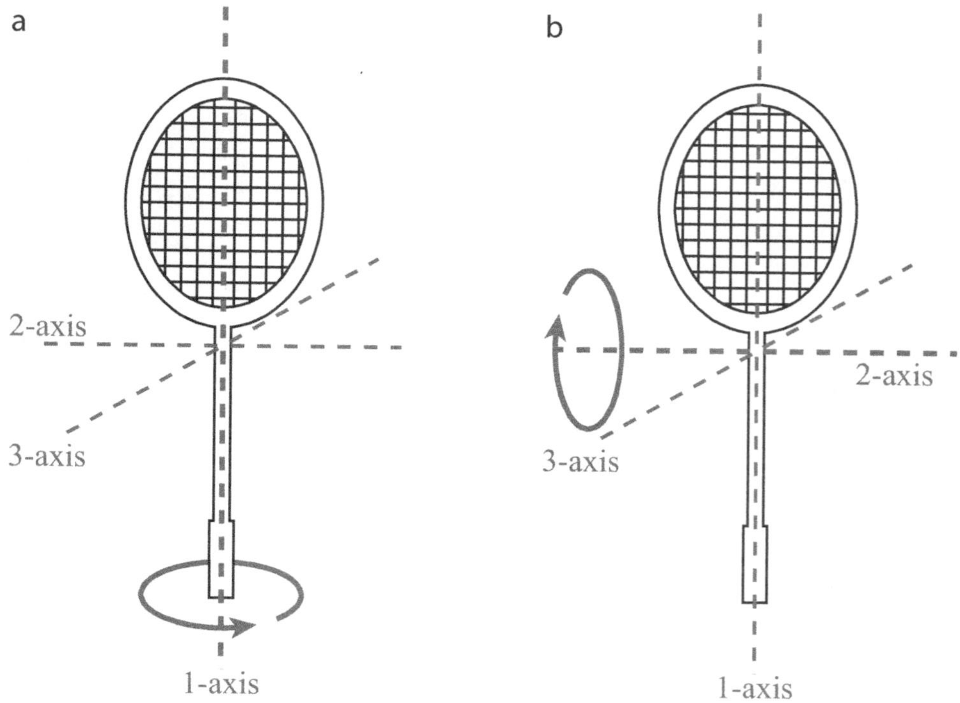

Figure 7.3: The three principal axes of a tennis racket.

7.4 PROBLEMS

7.1 Two Delft students wish to re-create Galilei's experiment dropping objects with different mass from a high tower. They use the tower of the Old Church in Delft, which, like the more famous one in Pisa supposedly used by Galilei, leans over somewhat. The tip of the tower is 75 m above street level, and 2.0 m removed from the vertical. The student who will drop the objects stands on the trans at 60 m. The base of the tower is a square of 10 × 10 m.

(a) How far from the base of the tower do you expect the stone to fall?

(b) The second student, who has done the same calculation you did in (a), has put a camera close to the floor aimed at the spot where the objects will drop. Surprisingly, in a test drop of a single stone, he observes that the actual position the stone hits the ground deviates from this spot. The student at the top however insists that she dropped the stone straight down from the tower trans, as agreed. The students therefore go back to their Mechanics books and realize that they forgot to account for the rotation of the Earth. Which of the (fictional) forces described in this section could cause the stone to deviate from its straight path down?

(c) Delft is located on the Northern hemisphere. In which direction will the trajectory of the stone be deflected?

(d) Delft is at 52.0°N. What is the magnitude of the deflection of the dropped stone on the ground? You may neglect air resistance in this calculation.

7.2 **Foucault's pendulum** A well-known (and conclusive) proof of the fact that the Earth is rotating is provided by a Foucault pendulum, first presented by French physicist Léon Foucault in 1851 (a replica of his device is on permanent exhibit in the Panthéon in Paris, as well as in many other science musea around the world, see figure 7.4). A key part of this pendulum is the way its pivot is constructed: it has to be rotationally symmetric and frictionless, so it can't exert any torques on the pendulum itself. Consequently, the plane in which the pendulum oscillates will remain unchanged[2], *even as* the Earth rotates. Therefore, for observers on Earth, the plane of the pendulum seems to rotate over time. To see how this works, consider putting this pendulum at the North pole. Then for an external observer, the plane of the pendulum stays

[2]With respect to the (relatively) fixed frame provided by a collection of distant stars.

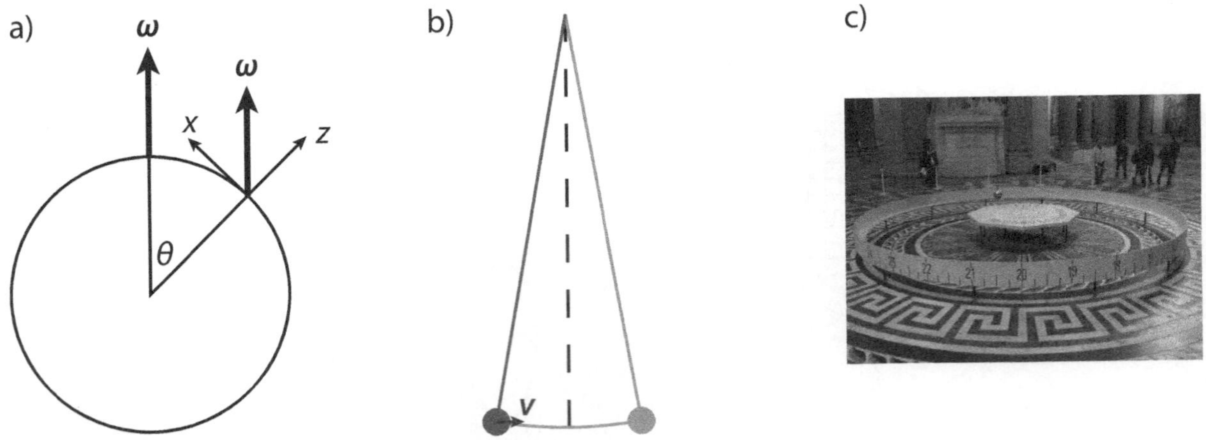

Figure 7.4: Foucault's pendulum. (a) Coordinates in Paris. (b) For a pendulum with a long string and small amplitude, the velocity of the bob will be almost horizontal. (c) The replica of Foucault's original pendulum at the Panthéon [19].

fixed (there are no forces acting on it), while the Earth (looking down on the North pole) rotates counter-clockwise; for an Earth-bound observer, the pendulum's plane thus seems to move clockwise (again as seen from the top), making one full revolution in one day.

Paris is not on the North pole, but it does lie on the Northern hemisphere, so the pendulum will still appear to rotate clockwise, just at a slower frequency. We'll calculate this precession frequency in this problem.

(a) First, we need the angular velocity in Paris, in a useful coordinate system. Define the $\hat{\boldsymbol{z}}$ axis as pointing upwards in Paris, and $\hat{\boldsymbol{x}}$ as the tangent to the planet due North (see figure 7.4a). Express $\boldsymbol{\omega}$ in these coordinates.

(b) If the pendulum has a very long string (the original Foucault one is 67 m) compared to its amplitude, the velocity v of the weight will be roughly in the horizontal direction, see figure 7.4b. Argue why, in this case, the component of $\boldsymbol{\omega}$ in the $\hat{\boldsymbol{x}}$ direction will not change the frequency at which the plane of the pendulum precesses.

(c) The pendulum's plane rotates with a (precession) frequency $\boldsymbol{\omega}_{\mathrm{P}} = \omega_{\mathrm{P}}\hat{\boldsymbol{z}}$ with respect to the Earth's frame fixed in Paris. This precession frequency must exactly compensate for the Earth's rotation in the frame of the pendulum (as in that frame, there are no forces acting on the pendulum, and thus its plane of oscillation stays fixed). Show that these considerations imply that $\omega_{\mathrm{P}} = -\omega\cos\theta$.

(d) Suppose that you enter the Panthéon at noon, and mark the direction in which the pendulum is oscillating. When you return an hour later, by how much will this plane have rotated? Will this be enough to be visible by eye? The Panthéon is at $48°50'46''$N (note that degrees, like hours, are divided in (arc)minutes and seconds that run up to 60, not 100).

7.3 An alternative way to show the effect of the rotation of the Earth involves only a smooth horizontal plane and a particle that can slide over it. Show that if the particle's velocity is v, its trajectory will be a circle with radius $r = v/2\Omega$, where Ω is the Earth's rotational velocity.

7.4 The centrifugal force emerges in a rotating coordinate frame, and famously causes the parabolic shape of the surface of water in a rotating bucket. As the centrifugal force is always perpendicular to the rotation axis, we can pick coordinates such that the rotation axis coincides with the z-axis, $\boldsymbol{\omega} = \omega\hat{\boldsymbol{z}}$, and we can express the centrifugal force in cylindrical coordinates as $\boldsymbol{F}_{\mathrm{cf}} = m\omega^2\rho\hat{\boldsymbol{\rho}}$ (equation 7.12).

We now consider a small volume of water at the rotating surface (in steady-state). There are two forces acting on this mass of water: gravity (pointing down, as always) and the centrifugal force, pointing outward, see figure 7.5. The resulting net force cannot have a component along the surface, as this would result in an acceleration of the water (and hence a water flow); therefore, the force must be perpendicular to the surface, and counterbalance the pressure in the water (just like it would for the flat surface of water in a bucket that is not rotating).

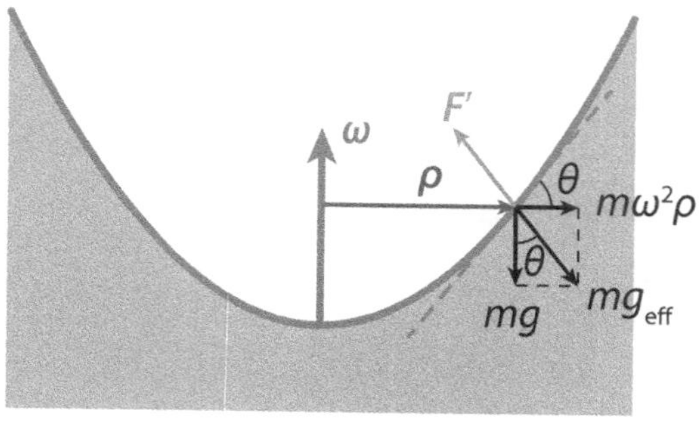

Figure 7.5: Parabolic water surface in a rotating bucket. The axis of rotation coincides with the axis of symmetry. For a small volume of water at the surface, the gravitational force and the centrifugal force add up to an effective gravitational force, which must be perpendicular to the water surface in steady-state. This net force is counterbalanced by the pressure in the water ($\boldsymbol{F}'$).

The gravitational and centrifugal force add up to what is known as an *effective gravity*, given by

$$\boldsymbol{g}_{\text{eff}} = -g\hat{\boldsymbol{z}} + \omega^2 \rho \hat{\boldsymbol{\rho}}. \tag{7.29}$$

(a) Find the angle θ the direction of the effective gravitational force makes with the vertical (see figure 7.5).

(b) If the gravitational force is to be perpendicular to the water surface, we must have

$$\frac{\mathrm{d}z}{\mathrm{d}\rho} = \tan\theta.$$

Integrate this equation to find $z(\rho)$ (and thus the shape of the surface).

(c) Find the potential energy corresponding to the effective gravitational force, $\boldsymbol{F}_{\text{eff}} = m\boldsymbol{g}_{\text{eff}}$.

(d) Argue why the potential energy must be constant on the water surface, and from this condition, again derive the shape of the water surface.

8

OSCILLATIONS

8.1 OSCILLATORY MOTION

8.1.1 HARMONIC OSCILLATOR

We've already encountered two examples of oscillatory motion - the rotational motion of chapter 5, and the mass-on-a-spring system in section 2.3 (see figure 1.1). The latter is the quintessential oscillator of physics, known as the *harmonic oscillator*. Recapping briefly, we get its equation of motion by considering a mass m that is being pulled on by a massless ideal spring of spring constant k. Equating the resulting spring force (Hooke's law) to the net force in Newton's second law of motion, we get:

$$m\ddot{x} = -kx. \tag{8.1}$$

The harmonic oscillator is characterized by its *natural frequency* ω_0:

$$\omega_0 = \sqrt{\frac{k}{m}}, \tag{8.2}$$

as follows readily by dimensional arguments (or, of course, by solving the differential equation). Because equation (8.1) is second-order, its solution has two unknowns; moreover, as it has to be minus its own derivative we readily see that it should be a linear combination of sines and cosines (for a formal derivation, see appendix A.3.2). We can write the solution in two different ways:

$$x(t) = x(0)\cos(\omega_0 t) + \frac{v(0)}{\omega_0}\sin(\omega_0 t), \tag{8.3}$$

$$= A\cos(\omega_0 t + \phi), \tag{8.4}$$

where the phase ϕ is given by $\tan\phi = -\frac{1}{\omega}\frac{v(0)}{x(0)}$ and the amplitude A by $A = \frac{x(0)}{\cos\phi}$. Unsurprisingly, as they are both simple periodic motions, there is a direct relationship between a harmonic oscillator with natural frequency ω_0, and a point on a disk rotating with uniform angular velocity ω_0 in the xy-plane - the motion of the harmonic oscillator is that of the disk projected on the x (or y) axis.

8.1.2 TORSIONAL OSCILLATOR

A torsional oscillator is the rotational analog of a harmonic oscillator - imagine a disk with moment of inertia I suspended by a massless, frictionless rope that has a torsional constant κ, i.e., the force to twist the rope is given by $F = -\kappa\theta$, with θ the twist angle. By invoking the rotational analog of Newton's second law of motion, equation (5.12), we readily find for the equation of motion of the torsional oscillator:

$$I\ddot{\theta} = -\kappa\theta, \tag{8.5}$$

so the torsional oscillator indeed is the exact rotational analog of the harmonic oscillator, and has a natural frequency of $\omega_0 = \sqrt{\kappa/I}$.

Christiaan Huygens (1629-1695) was a Dutch physicist and astronomer, and one of the major figures in the scientific revolution. Huygens invented the pendulum clock in 1656, which revolutionized timekeeping and remained the most accurate clock for 300 years. Huygens was also the first to cast the laws of physics in mathematical form, writing down an early (quadratic) version of Newton's second law of motion, the equation for the centripetal force (eq. 5.10), and the correct form of the laws of elastic collisions (section 4.7). Observing two pendulum clocks on the same wall, Huygens observed that they synchronized (see section 8.4). Huygens' study of optics led him to formulate the wave theory of light, which can correctly predict light diffraction. In astronomy, he discovered the first feature on the surface of Mars, the largest moon of Saturn (Titan), and that the previously observed 'shape changes' of Saturn were due to the presence of its rings. The Huygens probe that landed on Titan in 2005 was very appropriately named in his honor.

Figure 8.1: 1671 portrait of Huygens by Caspar Netscher [23].

8.1.3 Pendulum

A pendulum is an object that is suspended on a horizontal peg through any point x_P but its center of mass x_{CM} (it won't oscillate if you pin it at the center of mass). If the center of mass of the pendulum is pulled sideways, gravity will exert a torque around the position of the peg, pulling the pendulum back down. If the distance between x_P and x_{CM} is L, and the line connecting them makes an angle θ with the vertical through x_P, then the torque exerted by gravity around x_P equals $-mgL\sin\theta$, where as usual m is the mass of the pendulum. Now again invoking equation 5.12, we can write for the equation of motion of the pendulum (with I its moment of inertia about x_P):

$$I\ddot{\theta} = -mgL\sin\theta. \tag{8.6}$$

Unfortunately we can't solve equation (8.6). For small angles however, we can Taylor-expand the sine, and write $\sin\theta \approx \theta$, which takes us back to the harmonic oscillator equation. From that we find that for this pendulum (called the *physical pendulum*), the natural frequency is $\omega_0 = \sqrt{mgL/I}$. For the special case that the pendulum consists of a mass m suspended on a massless rope of length L (the *simple pendulum*), we have $I = mL^2$ and thus $\omega_0 = \sqrt{g/L}$.

8.1.4 Oscillations in a potential energy landscape

The potential energy associated with a mass on a spring has a very simple form: $U_s(x) = \frac{1}{2}kx^2$ (see equation 3.15). The potential energy landscape of a harmonic oscillator thus has the shape of a parabola. Now that's a shape that we encounter very often: the shape of pretty much every landscape about a minimum closely resembles a parabola[1]. To see why this is the case, simply Taylor-expand the potential energy about a minimum at x_0: because the function has a minimum at x_0, $U'(x_0) = 0$, and the Taylor expansion gives

$$U(x) = U(x_0) + \frac{1}{2}U''(x_0)x^2 + \mathcal{O}(x^3). \tag{8.7}$$

Around a minimum in the potential energy, any potential energy thus resembles that of a harmonic oscillator. Any particle placed in such a potential energy landscape close to a minimum (i.e., a particle on which a force acts close to the point where the force vanishes) will therefore tend to oscillate. By comparing equation (8.7) with the potential energy of the harmonic oscillator, we can immediately read off that the resulting oscillatory motion is identical to that of a harmonic oscillator with spring constant $k = U''(x_0)$. A particle released close to a minimum of the potential energy will thus oscillate with a frequency $\omega = \sqrt{U''(x_0)/m}$.

8.2 Damped harmonic oscillator

So far we've disregarded damping on our harmonic oscillators, which is of course not very realistic. The main source of damping for a mass on a spring is due to drag of the mass when it moves through air (or any fluid, either gas or liquid). For relatively low velocities, drag forces on an object scale linearly with the object's velocity, as illustrated by Stokes' law (equation 2.11). For an object of arbitrary shape moving through an

[1]The only exception being functions of the form x^{2n} for $n > 1$.

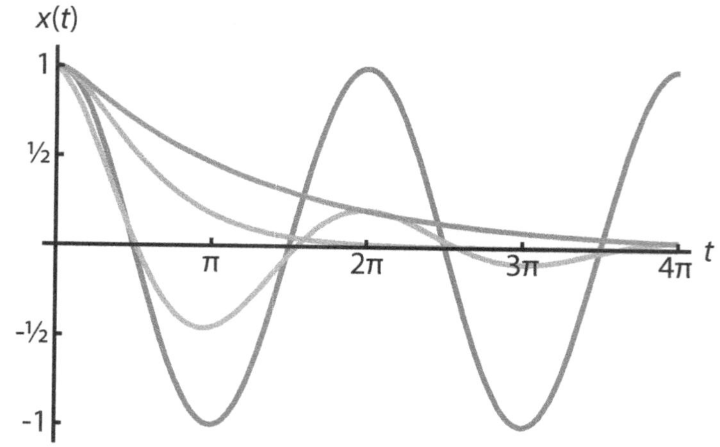

Figure 8.2: Position as function of time for four types of oscillation: undamped ($\zeta = 0$, blue), underdamped ($0 < \zeta < 1$, orange), critically damped ($\zeta = 1$, green) and overdamped ($\zeta > 1$, red). In all cases the initial conditions are $x(0) = 1$ and $v(0) = 0$.

arbitrary fluid we'll write $F_{\text{drag}} = -\gamma \dot{x}$, with γ the drag coefficient, and of course opposing the direction of motion. Adding this to the spring force gives for the equation of motion of the *damped harmonic oscillator*:

$$m\ddot{x} = -\gamma \dot{x} - kx. \tag{8.8}$$

We now have two numbers that determine the motion: the undamped frequency $\omega_0 = \sqrt{k/m}$ and the damping ratio $\zeta = \gamma/2\sqrt{mk}$. In terms of these parameters, we can rewrite equation (8.8) as:

$$\ddot{x} + 2\zeta\omega_0\dot{x} + \omega_0^2 x = 0. \tag{8.9}$$

The solution of equation (8.9) depends strongly on the value of ζ, see figure 8.2. We can find it[2] by substituting the Ansatz $x(t) = e^{\lambda t}$, which gives a characteristic equation for λ:

$$\lambda^2 + 2\zeta\omega_0\lambda + \omega_0^2 = 0, \tag{8.10}$$

so

$$\lambda = -\zeta\omega_0 \pm \omega_0\sqrt{\zeta^2 - 1}. \tag{8.11}$$

For $\zeta < 1$, there are two complex solutions for λ, and we find that $x(t)$ undergoes an oscillation with an exponentially decreasing amplitude:

$$x(t) = e^{-\zeta\omega_0 t}[A\cos(\omega_d t) + B\sin(\omega_d t)], \tag{8.12}$$

where $\omega_d = \omega_0\sqrt{1-\zeta^2}$ and A and B follow from the initial conditions. Because there is still an oscillation, this type of motion is called *underdamped.* In contrast, if $\zeta > 1$, the roots $\lambda_\pm$ in equation (8.11) are real, and we get qualitatively different, *overdamped* behavior, in which x returns to 0 with an exponential decay without any oscillations:

$$x(t) = Ae^{\lambda_+ t} + Be^{\lambda_- t} = e^{-\zeta\omega_0 t}\left[Ae^{\Omega t} + Be^{-\Omega t}\right], \tag{8.13}$$

where $\Omega = \omega_0\sqrt{\zeta^2 - 1}$. Naturally the boundary case is when $\zeta = 1$, which is a *critically damped* oscillator - the fastest return to 0 without oscillations. Because in this case equation (8.11) only has one root, we again get a qualitatively different solution:

$$x(t) = (A + Bt)e^{-\omega_0 t}. \tag{8.14}$$

The three different cases and the undamped oscillation are shown in figure 8.2.

[2]See appendix A.3.2 for the mathematical details on how to solve general equations of this type.

8.3 Driven harmonic oscillator

A mass on a spring, displaced out of its equilibrium position, will oscillate about that equilibrium for all time if undamped, or relax towards that equilibrium when damped. Its amplitude will remain constant in the first case, and decrease monotonically in the second. However, if we give the mass a periodic small push at the right moment in its oscillation cycle, its amplitude can increase, and even diverge. To see how this works we study the *driven oscillator*, where we apply a periodic driving force $F_D(t) = F_D\cos(\omega_D t) = \frac{1}{2}F_D\left(e^{i\omega_D t} + e^{-i\omega_D t}\right)$. Adding this driving force to the equation of motion (8.8) of a damped harmonic oscillator, we obtain:

$$\ddot{x} + 2\omega_0\zeta\dot{x} + \omega_0^2 x = \frac{F_D}{2m}\left(e^{i\omega_D t} + e^{-i\omega_D t}\right). \tag{8.15}$$

We already know the homogeneous solution to equation (8.15) - that's just the damped oscillator again, so depending on the value of ζ, we get one of the three possible solutions of the previous section. To find a particular solution, we first note that we can split the driving term in two - if we have a particular solution for each of the oscillating exponentials, we can simply add them. Also, these exponentials themselves look very similar to the underdamped solutions, so they may make a good guess for a particular solution. For a right-hand side of $(F_D/2m)e^{\pm i\omega_D t}$ we therefore try $x_p = Ae^{\pm i\omega_D t}$. Substituting this into equation (8.15) with the appropriate right-hand side, we get:

$$A\left(-\omega_D^2 \pm 2i\omega_0\zeta\omega_D + \omega_0^2\right)e^{\pm i\omega_D t} = \frac{F_D}{2m}e^{\pm i\omega_D t}, \tag{8.16}$$

so we find that we have indeed a solution if the amplitude is given by

$$A(\omega_D) = \frac{F_D}{2m\left(\omega_0^2 \pm 2i\omega_0\zeta\omega_D - \omega_D^2\right)}. \tag{8.17}$$

The full particular solution of equation (8.15) is then given by

$$\begin{aligned} x_p(t) &= \frac{F_D}{2m}\left[\frac{e^{i\omega_D t}}{\omega_0^2 + 2i\omega_0\zeta\omega_D - \omega_D^2} + \frac{e^{-i\omega_D t}}{\omega_0^2 - 2i\omega_0\zeta\omega_D - \omega_D^2}\right] \\ &= \frac{F_D}{m}\left[\frac{(\omega_0^2 - \omega_D^2)\cos(\omega_D t) + 2\omega_0\zeta\omega_D\sin(\omega_D t)}{(\omega_0^2 - \omega_D^2)^2 + 4\omega_0^2\zeta^2\omega_D^2}\right] \\ &= \frac{F_D}{mR(\omega_D)}\cos\left(\omega_D t - \phi(\omega_D)\right) \end{aligned} \tag{8.18}$$

where the factor $R(\omega_D)$ in the amplitude is defined by

$$R^2(\omega_D) = (\omega_0^2 - \omega_D^2)^2 + 4\omega_0^2\zeta^2\omega_D^2, \tag{8.19}$$

and the phase $\phi(\omega_D)$ by $\cos\phi = (\omega_0^2 - \omega_D^2)/R(\omega_D)$, $\sin\phi = 2\omega_0\zeta\omega_D/R(\omega_D)$, so

$$\tan(\phi(\omega_D)) = \frac{2\omega_0\zeta\omega_D}{(\omega_0^2 - \omega_D^2)}. \tag{8.20}$$

Resonance, a large response of the harmonic oscillator to a small driving force, occurs when $x_p(t)$ blows up, or $R(\omega_D)$ goes to zero. That does not always happen, but $R(\omega_D)$ can reach a minimum at which the amplitude becomes large:

$$0 = \frac{\mathrm{d}R^2}{\mathrm{d}\omega_D} = -4(\omega_0^2 - \omega_D^2)\omega_D + 8\omega_0^2\zeta^2\omega_D, \tag{8.21}$$

which is at

$$\omega_D^2 = \omega_0^2 - 2\omega_0^2\zeta^2, \tag{8.22}$$

or $\omega_D \simeq \omega_0$ if the damping factor ζ is small. Note that in this same limit (small ζ), we find that when $\omega_D \simeq \omega_0$, $\tan\phi \to \infty$, so $\phi \to \pi/2$. Therefore, in this case the driving happens out of phase with the response, that is to say, you push hardest when the mass is at its point of maximum speed, increasing that speed even further, and leading to an increase in amplitude. In practice, this is what kids do when they sit on a swing: they fling back their legs when they go through the lowest point (maximum speed) going backwards, and fling their legs forward at the same point when going forwards, increasing their speed and thus amplitude.

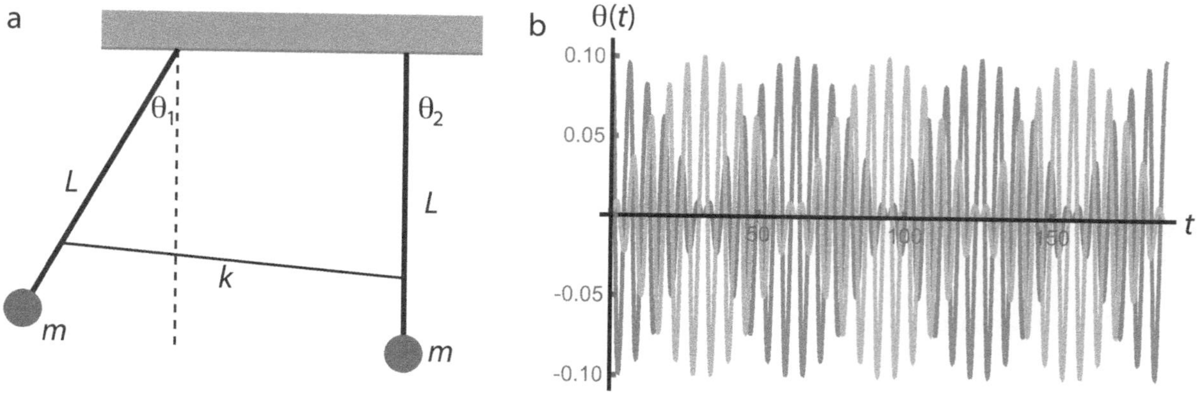

Figure 8.3: Motion of two coupled pendulums. (a) Sketch of the setup. Two identical pendulums of length L and mass m are connected through a weak spring of spring constant k. As our initial condition we choose both pendulums at rest, with the right one in its equilibrium position and the left one given a finite amplitude. (b) Resulting motion of the two pendulums: left (blue) and right (orange).

8.4 COUPLED OSCILLATORS

8.4.1 TWO COUPLED PENDULUMS

A beautiful demonstration of how energy can be transferred from one oscillator to another is provided by two weakly coupled pendulums. Imagine we have two identical pendulums of length L and mass m, which are connected by a weak spring with spring constant k (figure 8.3a). The equation of motion of the combined system is then given by:

$$L\ddot{\theta}_1 = -g\sin\theta_1 - kL(\sin\theta_1 - \sin\theta_2), \tag{8.23a}$$
$$L\ddot{\theta}_2 = -g\sin\theta_2 + kL(\sin\theta_1 - \sin\theta_2). \tag{8.23b}$$

We will once again use the small angle expansion in which we can approximate $\sin\theta \approx \theta$, and identify $\omega_0 = \sqrt{g/L}$ as the frequency of each of the (uncoupled) pendulums. Equations (8.23) then become

$$\ddot{\theta}_1 = -\omega_0^2\theta_1 - k\theta_1 + k\theta_2, \tag{8.24a}$$
$$\ddot{\theta}_2 = -\omega_0^2\theta_2 + k\theta_1 - k\theta_2. \tag{8.24b}$$

We can solve the system of coupled equations (8.24) easily by introducing two new variables: $\alpha = \theta_1 + \theta_2$ and $\beta = \theta_1 - \theta_2$, which gives us two uncoupled equations:

$$\ddot{\alpha} = -\omega_0^2\alpha, \tag{8.25a}$$
$$\ddot{\beta} = -\omega_0^2\beta - 2k\beta = -(\omega')^2\beta, \tag{8.25b}$$

where $(\omega')^2 = \omega_0^2 + 2k$ or $\omega' = \sqrt{2k + g/L}$. Since equations (8.25a) and (8.25b) are simply the equations of harmonic oscillators, we can write down their solutions immediately:

$$\alpha(t) = A\cos(\omega_0 t + \phi_0), \tag{8.26a}$$
$$\beta(t) = B\cos(\omega' t + \phi'). \tag{8.26b}$$

Converting back to the original variables θ_1 and θ_2 is also straightforward, and gives

$$\theta_1 = \frac{1}{2}(\alpha + \beta) = \frac{A}{2}\cos(\omega_0 t + \phi_0) + \frac{B}{2}\cos(\omega' t + \phi'), \tag{8.27a}$$
$$\theta_2 = \frac{1}{2}(\alpha - \beta) = \frac{A}{2}\cos(\omega_0 t + \phi_0) - \frac{B}{2}\cos(\omega' t + \phi'). \tag{8.27b}$$

Let's put in some specific initial conditions: we leave pendulum number 2 at rest in its equilibrium position ($\theta_2(0) = \dot{\theta}_2(0) = 0$ and give pendulum number 1 a finite amplitude but also release it at rest ($\theta_1(0) = \theta_0$,

$\dot{\theta}_1(0) = 0$). Working out the four unknowns (A, B, ϕ_0 and ϕ') is straightforward, and we get:

$$\theta_1 = \frac{\theta_0}{2}\cos(\omega_0 t) + \frac{\theta_0}{2}\cos(\omega' t) = \theta_0 \cos\left(\frac{\omega_0+\omega'}{2}t\right)\cos\left(\frac{\omega_0-\omega'}{2}t\right), \tag{8.28a}$$

$$\theta_2 = \frac{\theta_0}{2}\cos(\omega_0 t) - \frac{\theta_0}{2}\cos(\omega' t) = \theta_0 \sin\left(\frac{\omega_0+\omega'}{2}t\right)\sin\left(\frac{\omega'-\omega_0}{2}t\right). \tag{8.28b}$$

The solution given by equations (8.28) is plotted in figure 8.3b. Note that the solutions have two frequencies (known as the *eigenfrequencies* of the system). The fast one, $\frac{1}{2}(\omega_0+\omega')$, which for a weak coupling constant k is very close to the eigenfrequency ω_0 of a single pendulum, is the frequency at which the pendulums oscillate. The do so in anti-phase, as expressed mathematically by the fact that one oscillation has a sine and the other a cosine (which is of course just a sine shifted over $\pi/2$). The second frequency, $\frac{1}{2}(\omega'-\omega_0)$ is much slower, and represents the frequency at which the two pendulums transfer energy to each other, through the spring that couples them. In figure 8.3b, it is the frequency of the envelope of the amplitude of the oscillation of either of the pendulums. All these phenomena will return in the next section, in the study of waves, which travel in a medium in which many oscillators are coupled to one another.

8.4.2 Normal modes

For a system with only two oscillators, the technique we used above for solving the system of coupled equations (8.23) is straightforward. It does however not generalize easily to systems with many oscillators. Instead, we can exploit the fact that the equations are linear and use techniques from linear algebra (as you may have guessed from the term eigenfrequency). We can rewrite equations (8.23) in matrix form:

$$\frac{d^2}{dt^2}\begin{pmatrix}\theta_1\\ \theta_2\end{pmatrix} = \begin{pmatrix}-(\omega_0^2+k) & k\\ k & -(\omega_0^2+k)\end{pmatrix}\begin{pmatrix}\theta_1\\ \theta_2\end{pmatrix}. \tag{8.29}$$

Equation (8.29) is a homogeneous, second-order differential equation with constant coefficients, strongly resembling the equation for a simple, one-dimensional harmonic oscillator. Consequently, we may expect the solutions to look similar as well, so we try our usual Ansatz:

$$\begin{pmatrix}\theta_1\\ \theta_2\end{pmatrix} = \begin{pmatrix}C_1\\ C_2\end{pmatrix}e^{i\omega t}, \tag{8.30}$$

where C_1 and C_2 are constants. Substituting (8.30) in (8.29) gives

$$\begin{pmatrix}\omega_0^2+k & -k\\ -k & \omega_0^2+k\end{pmatrix}\begin{pmatrix}C_1\\ C_2\end{pmatrix} = \omega^2\begin{pmatrix}C_1\\ C_2\end{pmatrix}, \tag{8.31}$$

which you hopefully recognize as an eigenvalue problem. Solving for the eigenvalues ω^2 gives:

$$(-\omega^2+\omega_0^2+k)^2 - k^2 = 0. \tag{8.32}$$

The solutions of equation (8.32) unsurprisingly reproduce the frequencies of the uncoupled equations in section 8.4.1:

$$\omega_+^2 = \omega_0^2, \quad \omega_-^2 = \omega_0^2 + 2k. \tag{8.33}$$

The eigenvectors of (8.31) are given by

$$\boldsymbol{C}_+ = \frac{1}{\sqrt{2}}\begin{pmatrix}1\\ 1\end{pmatrix} \quad \text{for} \quad \omega_+ \quad \text{and} \quad \boldsymbol{C}_- = \frac{1}{\sqrt{2}}\begin{pmatrix}1\\ -1\end{pmatrix} \quad \text{for} \quad \omega_-. \tag{8.34}$$

Note that the eigenvectors are orthogonal; this is a general property of the eigenvectors of symmetric matrices. Each eigenvector corresponds to a possible steady-state of motion of the system; these states are known as the *normal modes* ('normal' referring to the orthogonality of the eigenvectors). We can now immediately write down the most general solution of equation (8.29) as a linear combination of the eigenmodes:

$$\begin{pmatrix}\theta_1(t)\\ \theta_2(t)\end{pmatrix} = \frac{A_+}{2}e^{i(\omega_+ t+\phi_+)}\begin{pmatrix}1\\ 1\end{pmatrix} + \frac{A_-}{2}e^{i(\omega_- t+\phi_-)}\begin{pmatrix}1\\ -1\end{pmatrix}, \tag{8.35}$$

where the amplitudes $A_\pm$ and phases $\phi_\pm$ are determined by the initial conditions.

Writing our system of equations in matrix form allows us to easily generalize both to asymmetric configurations (see problem 8.4.7) and to systems with many coupled oscillators. An important example of the latter case is the study of vibrations in solids. Atoms or ions in solids typically form a crystal lattice, that can be modeled as a large number of masses coupled by springs. Such crystals can have complicated vibrational properties, that can be analyzed in terms of its normal modes. In particular the modes with a low energy can typically be accessed easily. They are known as *phonons*, and correspond to sound waves in the solid.

8.5 PROBLEMS

8.1 An object undergoes simple harmonic motion of amplitude A and angular frequency ω about the equilibrium point $x = 0$. Find the speed v of the object in terms of A, ω, and x. *Hint*: use conservation of energy.

8.2 A disk of radius R and mass M is suspended from a pivot somewhere between its center and its edge, see figure below. For what pivot point (i.e., which distance d) will the period of this physical pendulum be a minimum (or equivalently the frequency a maximum)? You may find one of the theorems we proved in chapter 5 useful in answering this question.

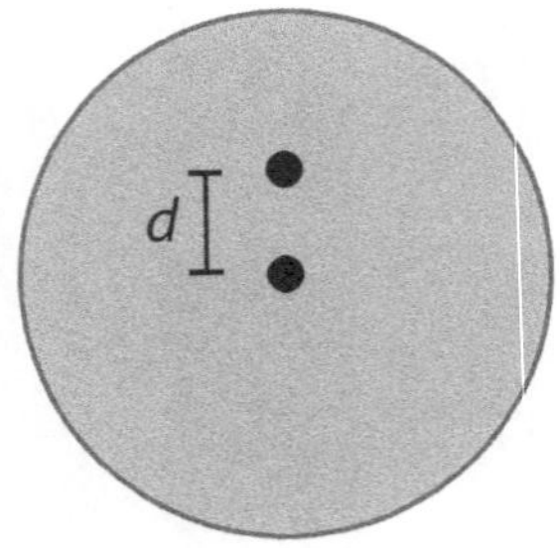

8.3 Figure 8.4 shows a common present-day seesaw design, also featured in problem 2.10. In addition to a beam with two seats, this seesaw also contains two identical springs (with spring constant 10 kN/m) that connect the beam to the ground. The distance between the pivot and each of the springs is 30.0 cm, the distance between the pivot and each of the seats is 1.50 m. Two children sit on the two seats. Both

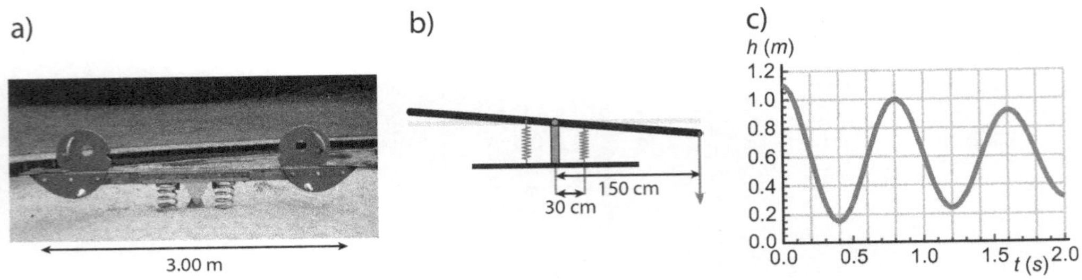

Figure 8.4: Seesaw with two springs.

children kick off against the ground a couple of time, putting the seesaw in an oscillating motion with an amplitude of 50.0 cm. At $t = 0$ the children stop kicking. The plot in figure c shows the height of one of the seats as a function of time afterwards.

(a) In what type of motion is the seesaw after the children stop kicking?

(b) You could model the seesaw with the two children as a simple mass-on-a-spring, with a spring constant twice that of the individual spring in the seesaw. Using the graph in figure c, estimate the effective mass of this system.

(c) After a while, the children resume kicking, slowly bringing their amplitude back up to 50.0 cm. Using the mass-spring system of (b), estimate the amount of energy the children have to put in per period to achieve this.

8.4 A block with mass $m_1 = 1.5$ kg block is supported by a frictionless surface and attached to a horizontal spring of constant $k = 22$ N/m, as shown in the figure. The block oscillates with an amplitude of 10.0 cm, executing a simple harmonic motion.

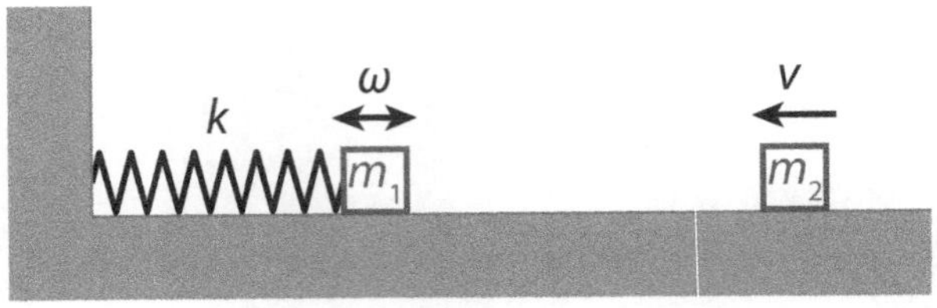

(a) Find the frequency ω of the oscillation of the block.

(b) Write down the equation for the position of the block as a function of time, $x(t)$, in such a form that it is at its rightmost position at $t = 0$.

A second block of mass 0.80 kg moves in from the right at 2.5 m/s and hits the first block at $t = 0$, i.e., when it is in its rightmost position. The two blocks then stick together and continue moving as one.

(c) Which quantity / quantities are conserved during the collision?

(d) Determine the frequency of the motion of the two blocks after the collision.

(e) Determine the amplitude of the motion of the two blocks after the collision.

8.5 Suppose you are stranded on an unknown planet with nothing but a physical pendulum and a stopwatch. You determined the properties of the pendulum back on Earth, and found $m = 2.0$ kg, $h = 0.50$ m and $I = 3.0\,\mathrm{kg \cdot m^2}$. Having nothing better to do, you measure the time it takes your pendulum to complete 50 cycles, and find that this time equals 170 s. Use this information to compute the value of the gravitational acceleration g on your new home world.

8.6 For a damped harmonic oscillator driven by a sinusoidal force (as in equation 8.15), find the average power dissipated per (driving) period. *Hint*: use $P = F \cdot v$.

8.7 Consider a system of two coupled harmonic oscillators, where one (with mass $2m$ and spring constant $2k$) is suspended from the ceiling, and the other (with mass m and spring constant k) is suspended from the first, as shown in the figure.

(a) Find the equation of motion of this system of coupled oscillators, and write it in matrix form. For each mass, use coordinates in which the zero is at the equilibrium position.

(b) Find the frequencies of the normal modes of this coupled system.

9

WAVES

In physics a *wave* is a disturbance or oscillation that travels through space accompanied by a transfer of energy, and may be propagated with little or no net motion of the medium involved. In this section we will consider *mechanical waves*, in which the particles in a material are oscillating. Examples are the waves in the sea, the wave in the crowd at a stadium, and sound. Later on we will encounter *electromagnetic waves* in which electric and magnetic fields are oscillating, and which can travel through vacuum. Examples are light and radio signals. In quantum mechanics, we will also encounter what are sometimes referred to as *matter waves*, where fundamental objects that we usually think of as particles, such as electrons and protons, can also be considered as waves. Finally, recently *gravitational waves* were discovered, which are vibrations of spacetime itself.

By observing a particle, we know in which direction it moves at any given time. However, as I just stated, the particles in a mechanical wave have no, or almost no, net motion as the wave passes. The wave does have a well-defined direction though: the direction in which energy is transferred. Some waves spread out uniformly, such as a sound wave emanating from a point source. Others are restricted in their motion by the properties of the material they travel in, such as a wave in a string, or by boundary conditions, such as the end of that string. For waves that move (predominantly) in one direction, we can distinguish two fundamental types, illustrated in figure 9.1. The first type is the case that the particles oscillate in the same direction as the wave is moving (figure 9.1a), which we call a *longitudinal* wave; sound is an example. The second case is that the particles oscillate in a direction perpendicular to the wave motion, which we call a *transverse* wave (figure 9.1b), of which the waves in a pond are an example.

9.1 SINUSOIDAL WAVES

Probably the simplest kind of wave is a transverse sinusoidal wave in a one-dimensional string. In such a wave each point of the string undergoes a harmonic oscillation. We will call the displacement from equilibrium u, then we can plot u as a function of position on the string at a given point in time, figure 9.2a, which is a snapshot of the wave. Alternatively, we can plot u as a function of time for a given point (with given position) on the string, figure 9.2b. Because the oscillation is harmonic, the displacement as a function of time is a sine function, with an amplitude (maximum displacement) A and a period (time between maxima) T.

By definition, each point of the string undergoing a sinusoidal wave undergoes a harmonic oscillation, so for each point we can write $u(t) = A\cos(\omega t + \phi)$ (equation 8.4), where as before $\omega = 2\pi / T$ is the (angular) frequency and ϕ the phase. Two neighboring points on the string are slightly out of phase - if the wave is traveling to the right, then your right-hand neighbor will reach maximum slightly later than you, and thus has a slightly larger phase. The difference in phase is directly proportional to the distance between two points, coming to a full 2π (which is of course equivalent to zero) after a distance λ, the *wavelength*. The wavelength is thus the distance between two successive points with the same phase, in particular between two maxima. In between these maxima, the phase runs over the full 2π, so the wave is also a sinusoid in space, with a 'spatial frequency' or *wavenumber* $k = 2\pi / \lambda$. Combining the dependencies on space and time in a single expression, we can write for the sinusoidal wave:

$$u(x, t) = A\cos(kx - \omega t). \tag{9.1}$$

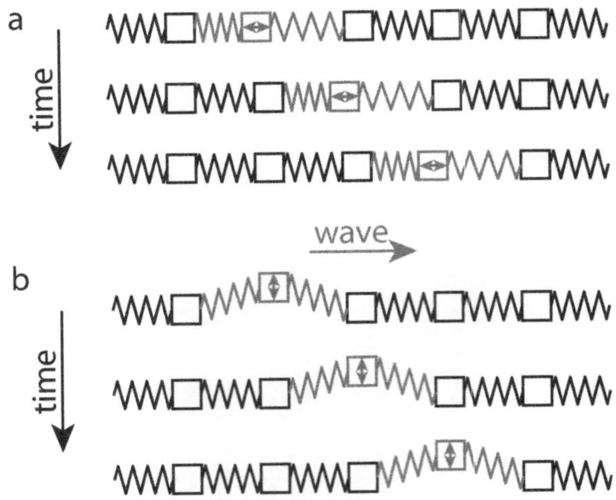

Figure 9.1: Two basic types of waves. (a) Longitudinal wave, where the oscillatory motion of the particles is in the same direction as that of the wave. (b) Transverse wave, where the oscillatory motion of the particles is perpendicular to that of the wave.

The speed of the wave is the distance the wave travels per unit time. A unit time for a wave is one period, as that is the time it takes the oscillation to return to its original point. The distance traveled in one period is one wavelength, as that is the distance between two maxima. The speed is therefore simply their ratio, which can also be expressed in terms of the wave number and frequency:

$$v_{\mathrm{w}} = \frac{\lambda}{T} = \frac{\omega}{k}. \tag{9.2}$$

9.2 THE WAVE EQUATION

As with all phenomena in classical mechanics, the motion of the particles in a wave, for instance the masses on springs in figure 9.1, are governed by Newton's laws of motion and the various force laws. In this section we will use these laws to derive an equation of motion for the wave itself, which applies quite generally to wave phenomena. To do so, consider a series of particles of equal mass m connected by springs of spring constant k, again as in figure 9.1a, and assume that at rest the distance between any two masses is h. Let the position of particle i be x, and u the distance that particle is away from its rest position; then $u = x_{\mathrm{rest}} - x$ is a function of both position x and time t. Suppose particle i has moved to the left, then it will feel a restoring force to the right due to two sources: the compressed spring on its left, and the extended spring on its right. The total force to the right is then given by:

$$\begin{aligned} F_i &= F_{i+1\to i} - F_{i-1\to i} \\ &= k\,[u(x+h,t) - u(x,t)] - k\,[u(x,t) - u(x-h,t)] \\ &= k\,[u(x+h,t) - 2u(x,t) + u(x-h,t)] \end{aligned} \tag{9.3}$$

Equation (9.3) gives the net force on particle i, which by Newton's second law of motion (equation 2.5) equals the particle's mass times its acceleration. The acceleration is the second time derivative of the position x, but since the equilibrium position is a constant, it is also the second time derivative of the distance from the equilibrium position $u(x,t)$, and we have:

$$F_{\mathrm{net}} = m\frac{\partial^2 u(x,t)}{\partial t^2} = k\,[u(x+h,t) - 2u(x,t) + u(x-h,t)]\,. \tag{9.4}$$

Equation (9.4) holds for particle i, but just as well for particle $i+1$, or $i-10$. We can get an equation for N particles by simply adding their individual equations, which we can do because these equations are linear. We thus find for a string of particles of length $L = Nh$ and total mass $M = Nm$:

$$\frac{\partial^2 u(x,t)}{\partial t^2} = \frac{KL^2}{M}\frac{u(x+h,t) - 2u(x,t) + u(x-h,t)}{h^2}. \tag{9.5}$$

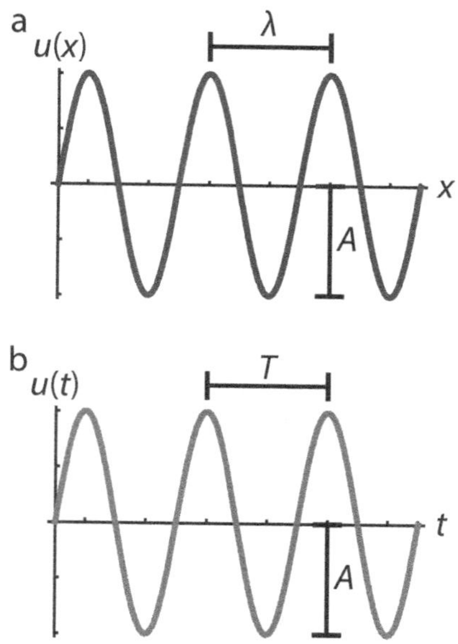

Figure 9.2: A sinusoidal transverse wave in space (a) and time (b). The distance between two successive maxima (or any two successive points with equal phase is the *wavelength* λ of the wave. The maximum displacement is the *amplitude* A, and the time it takes a single point to go through a full oscillation is the *period* T.

Here $K = k/N$ is the effective spring constant of the N springs in series. Now take a close look at the fraction on the right hand side of equation (9.5): if we take the limit $h \to 0$, this is the second derivative of $u(x,t)$ with respect to x. However, taking h to zero also takes L to zero - which we can counteract by simultaneously taking $N \to \infty$, in such a way that their product L remains the same. What we end up with is a string of infinitely many particles connected by infinitely many springs - so a continuum of particles and springs, for which the equation of motion is given by the *wave equation*:

$$\frac{\partial^2 u(x,t)}{\partial t^2} = v_{\mathrm{w}}^2 \frac{\partial^2 u(x,t)}{\partial x^2}. \tag{9.6}$$

In equation (9.6), $v_{\mathrm{w}} = \sqrt{KL^2/M}$ (sometimes also denoted by c) is the wave speed.

For a wave in a taut string, the one-dimensional description is accurate, and we can relate our quantities K, L and M to more familiar properties of the string: its tension $T = KL$ with the dimension of a force (this is simply Hooke's law again) and its mass per unit length $\mu = M/L$, so we get

$$v_{\mathrm{string}} = \sqrt{\frac{T}{\mu}}. \tag{9.7}$$

In two or three dimensions, the spatial derivative in equation (9.6) becomes a Laplacian operator, and the wave equation is given by:

$$\frac{\partial^2 u(\boldsymbol{x},t)}{\partial t^2} = v_{\mathrm{w}}^2 \nabla^2 u(\boldsymbol{x},t) \tag{9.8}$$

As can be readily seen by writing equation (9.8) in terms of spherical coordinates, if the wave is radial (i.e., only depends on the distance to the source r, and not on the angle), the quantity $ru(r)$ obeys the one-dimensional wave equation, so we can write down the equation for $u(r)$ immediately. An important application are sound waves, which spread uniformly in a uniform medium. To find their speed, we characterize the medium in a similar fashion as we did for the string: we take the mass per unit volume, which is simply the density ρ, and the medium's *bulk modulus*, which is a measure for the medium's resistance to compression (i.e., a kind of three-dimensional analog of the spring constant), defined as:

$$B = -V\frac{\mathrm{d}p}{\mathrm{d}V} = \rho\frac{\mathrm{d}p}{\mathrm{d}\rho}, \tag{9.9}$$

where p is the pressure (force per unit area) and V the volume. The bulk modulus is also sometimes denoted as K. The dimensions of the bulk modulus are those of a pressure, or force per unit area, and those of the density are mass per unit volume, so their ratio has the dimension of a speed squared, and the speed of sound is given by:

$$v_{\text{sound}} = \sqrt{\frac{B}{\rho}} \tag{9.10}$$

Equation (9.8) describes a wave characterized by a one-dimensional displacement (either longitudinal or transverse) in three dimensions. In general a wave can have components of both, and the displacement itself becomes a vector quantity, $\boldsymbol{u}(\boldsymbol{x}, t)$. In that case the three-dimensional wave equation takes on a more complex form:

$$\rho \frac{\partial^2 \boldsymbol{u}(\boldsymbol{x}, t)}{\partial t^2} = \boldsymbol{f} + (B + \frac{4}{3}G)\nabla(\nabla \cdot \boldsymbol{u}(\boldsymbol{x}, t)) - G\nabla \times (\nabla \times \boldsymbol{u}(\boldsymbol{x}, t)), \tag{9.11}$$

where $\boldsymbol{f}$ is the driving force (per unit volume), B again the bulk modulus, and G the material's *shear modulus*. Equation (9.11) is used for the description of seismic waves in the Earth and the ultrasonic waves with which solid materials are probed for defects.

9.3 SOLUTION OF THE ONE-DIMENSIONAL WAVE EQUATION

The one-dimensional wave equation (9.6) has a surprisingly generic solution, due to the fact that it contains second derivatives in both space and time. As you can readily see by inspection, the function $q(x, t) = x - v_{\mathrm{w}} t$ is a solution, as is the same function with a plus instead of a minus sign. These functions represent waves traveling to the right (minus) or left (plus) at speed v_{w}. However, the shape of the wave does not matter - any function $F(q) = F(x - v_{\mathrm{w}} t)$ is a solution of (9.6), as is any function $G(x + v_{\mathrm{w}} t)$, and the general solution is the sum of these:

$$u(x, t) = F(x - v_{\mathrm{w}} t) + G(x + v_{\mathrm{w}} t). \tag{9.12}$$

To find a specific solution, we need to look at the *initial conditions* of the wave, i.e., the conditions at $t = 0$. Because the wave equation is second order in time, we need to specify both the initial displacement and the displacement's initial velocity, which can be functions of the position. For the most general case we write:

$$u(x,0) = u_0(x), \tag{9.13}$$

$$\dot{u}(x,0) = v_0(x). \tag{9.14}$$

The resulting solution of the one-dimensional wave equation is known as *d'Alembert's equation*:

$$u(x, t) = \frac{1}{2}(u_0(x - v_{\mathrm{w}} t) + u_0(x + v_{\mathrm{w}} t)) + \frac{1}{2 v_{\mathrm{w}}} \int_{x - v_{\mathrm{w}} t}^{x + v_{\mathrm{w}} t} v(y)\,\mathrm{d}y. \tag{9.15}$$

9.4 WAVE SUPERPOSITION

The wave equation (9.6) is linear in the function we're interested in, the displacement $u(x, t)$. This simple mathematical statement has important consequences, because it means that if we know any set of solutions, we can create more solutions by making linear combinations of them - so if $u_1(x, t)$ and $u_2(x, t)$ are solutions, then so are $au_1(x, t) + bu_2(x, t)$ for any choice of a and b. In physics, this useful property of linear differential equations is known as the *principle of superposition*. Thanks to this principle, we can study how different waves interact with each other without having to do (much) extra math.

To illustrate, let us consider two one-dimensional waves traveling in opposite directions, figure 9.3. As long as the waves do not overlap, the oscillation of any given particle is due to only one wave, and there is no interaction. However, as soon as the waves start overlapping, the oscillations add up, which leads to *interference*. At some points, the two oscillations will be in phase, resulting in a much larger oscillation amplitude, which we call constructive interference (figure 9.3b). At other points, the two oscillations will be out of phase, resulting in a much smaller, or even vanishing oscillation amplitude, which we call destructive interference (figure 9.3c). However, the waves themselves remain unaffected, and transmit right through each other, continuing their path as if nothing had happened (figure 9.3d).

Waves reaching the end of a string, or edge of a pond, or any type of boundary, will not simply disappear. Remember, waves carry energy, and that energy is conserved, so it has to go somewhere once the wave reaches the boundary. If there's nothing at the boundary, the waves are reflected back into the material. This happens in two cases: a (perfectly) fixed boundary, and a (perfectly) free boundary; in other cases, some of the energy may be transmitted to material on the other side of the boundary (starting a new wave there),

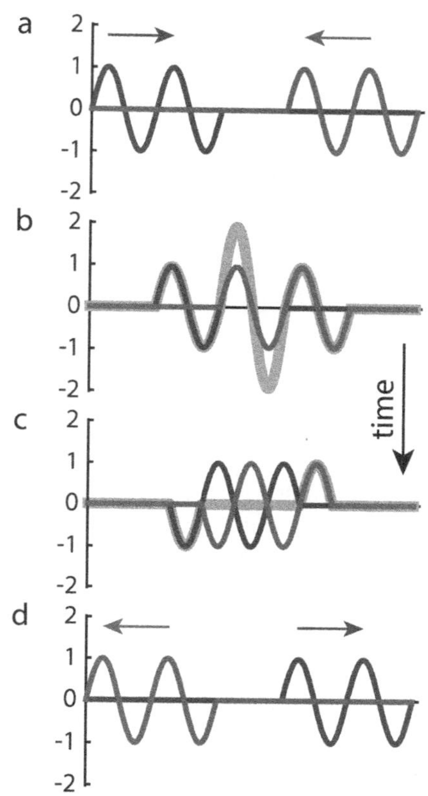

Figure 9.3: Two interacting wave packets. The sequence of images shows four snapshots. The blue wave is traveling to the right, the red wave to the left (a). When the waves overlap, the total displacement of the particle is given by the sum of the displacements due to both waves, shown in green. This can lead to both *constructive interference* (b), when the two waves are in phase, and *destructive interference* (c), when their phases are opposite. The waves themselves are not affected by the interaction and afterwards travel on as if nothing has happened (d).

whereas the remainder is reflected back into the original material with less energy, resulting in a smaller wave amplitude. A reflected wave travels in the opposite direction to the original wave, so it can interfere with itself. In fact, this interference is a crucial point for being able to meet the boundary conditions. A fixed boundary cannot move, so there must be destructive interference keeping the amplitude there zero at all times - so it follows that a wave reflecting on a fixed boundary undergoes a π phase shift. Free boundaries on the other hand are perfectly free to move, so there is nothing holding it back from reaching the maximal displacement that can be achieved by constructive interference, and the wave reflects without a phase shift.

If you put boundaries on both ends of a string, the wave keeps reflecting back and forth, continuously interfering with itself. To find the resulting shape of the string, we're going to use the principle of superposition for a simple sinusoidal wave. Let $u_1(x,t) = A\cos(kx-\omega t)$ be the part of the wave traveling to the right, and $u_2(x,t) = -A\cos(kx+\omega t)$ be the part traveling to the left. Note the differences: the waves have opposite signs for their speeds, and opposite signs for their displacements, the latter because of the π phase shift (we could also write $u_2(x,t) = A\cos(kx+\omega t+\pi)$). The shape of the string is now simply the sum of these two waves:

$$\begin{aligned} u(x,t) &= u_1(x,t)+u_2(x,t) = A[\cos(kx-\omega t)-\cos(kx+\omega t)] \\ &= 2A\sin(kx)\sin(\omega t). \end{aligned} \tag{9.16}$$

Equation (9.16) tells us that for a self-interfering wave, the wave no longer moves - instead, each point simply oscillates with frequency ω at a position-dependent amplitude $2A\sin(kx)$. We call such a wave a *standing wave*. Standing waves are very common - you'll get one every time you'll touch the string of a guitar or violin. Naturally, they are not restricted to one-dimensional systems - the skin of a drum, constrained at the drum's edge, is put in a standing wave every time someone hits it.

Equation (9.16) describes the shape of a standing wave on a string clamped at both ends. If the string has length L, then by the nature of the boundary conditions, we must have $u(0,t) = u(L,t) = 0$ for all t. The first condition follows for free (which is of course just due to a good choice of coordinates), but the second puts a constraint on our wave. The displacement can only be zero at all times if the amplitude is identically zero, so we demand that $\sin(kL) = 0$, or $L = m\pi/k = m\lambda/2$, where m is any positive integer. There are thus infinitely many allowed standing waves, but they are characterized by a discrete number. The allowed waves are known as *modes*, and the associated number m is the *mode number*. The simplest wave, with the lowest possible value, $m = 1$, is known as the *fundamental mode*. In the fundamental mode, the oscillation of the string has nonzero amplitude everywhere but at the fixed ends; for higher modes, there are also points in between that have zero amplitude, which are known as *nodes*; points where the amplitude is maximum are sometimes referred to as *antinodes*.

A discrete spectrum of allowed solutions, characterized by integer numbers, does not only appear in standing mechanical waves, but is also a fundamental aspect of quantum mechanics.

9.5 AMPLITUDE MODULATION

So far, we've mostly considered simple sinusoidal waves with fixed amplitudes. However, the general solution to the wave equation allows for many more interesting wave shapes. An important, and often encountered one is where the wave itself is used as the medium, by changing the amplitude over time:

$$u(x,t) = A(x,t)\cos(kx-\omega t). \tag{9.17}$$

The wave now consists of two waves: the carrier wave, which travels with the *phase velocity* $v_w = \omega/k$, and the *envelope*, which travels with the *group velocity* v_g. An illustration of a modulated wave is shown in figure 9.5. In the common case that the group velocity is independent of the wavelength of the carrier wave, we can rewrite (9.17) to reflect the fact that the amplitude is now also a wave, with speed v_g:

$$u(x,t) = A(x-v_g t)\cos(kx-\omega t). \tag{9.18}$$

9.6 SOUND WAVES

So far, we mostly considered transversal waves, which include waves in strings and waves on the surface of a pond, and are easily visualized. Longitudinal waves, on the other hand, are somewhat harder to draw, but easily heard - as sound is the prime example of a longitudinal wave. Other examples include (some forms of) seismic waves and ultrasound. Many people simply lump all of these together, and use the terms 'sound waves' and 'longitudinal waves' as synonyms.

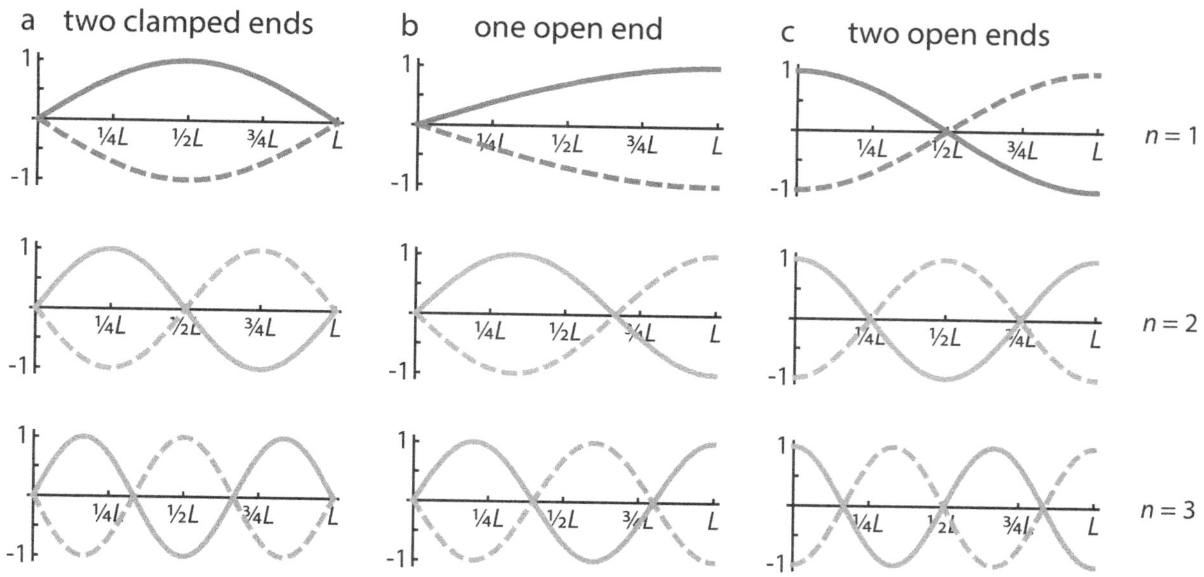

Figure 9.4: Snapshots of the wave patterns for the first three modes of a standing wave in (a) a string with two clamped ends, (b) a string with one clamped end, or a tube with one closed and one open end, and (c) a tube with two open ends. In each case, the plot shows the amplitude of the standing wave, as a fraction of the amplitude of the traveling wave. At nodes, the amplitude of the standing wave vanishes: destructive interference causes these points to always stand still. At antinodes, the amplitude of the standing wave is maximal, and equals that of the traveling wave. Note that (a) and (b) will occur for transversal waves, whereas (b) and (c) will occur for sound waves. For cases (a) and (c) allowed wavelengths are $\lambda = 2L/n$, whereas for case (b), allowed wavelengths are $\lambda = 4L/(2n-1)$.

We already touched upon the speed of sound waves in section 9.2 (equation 9.10). This speed indicates how fast the *wavefronts* of a sound wave travel; a wavefront is defined as the surface (in three dimensions) where all points have the same phase[1]. To visualize a sound wave we draw a succession of wavefronts one wavelength (or one period) apart. Simple examples include a point source (generating spherical wavefronts) and a planar wave, in which all wave fronts are parallel planes (see figure 9.6). The (local) direction of propagation of the wave is the direction perpendicular to the wavefronts, sometimes depicted by a ray.

Like transversal waves, longitudinal waves exhibit interference, both with other waves they encounter, and by their own reflections. There can therefore be traveling and standing sound waves. Unlike transversal waves on strings however, longitudinal sound waves are typically created in tubes that are open on either one or both ends. A closed end represents a fixed point, just like the fixed end of a string does, resulting in a stringent boundary condition: the interference between the incoming and outgoing wave must be such that the net displacement at the closed end is zero. An open end corresponds to a transversal wave in a string that is not clamped. As the string in that case is free to move, its maximum displacement will equal the amplitude of the wave. In other words, while a closed end corresponds to a node, the open end corresponds to an antinode. For a string (or pipe) with one clamped / closed and one open end, the wavelength of the lowest-order mode (known as the fundamental mode or first harmonic) therefore equals four times the length L of the string / pipe. The next mode (second harmonic) will have a wavelength of $4L/3$, and so on (see figure 9.4b), resulting in $\lambda = 4L/(2n-1)$ for the nth mode. For a tube with two open ends, the fundamental mode is the inverse of that of a string with two clamped ends - so two antinodes at the ends, and a node in the middle (figure 9.4c). Like for a clamped string, the allowed wavelengths are therefore $\lambda = 2L/n$.

9.7 THE DOPPLER EFFECT

The *Doppler effect* is a physical phenomenon that most people have experienced many times: when a moving source of sound (say an ambulance, or more exactly its siren) is approaching you, its pitch sounds noticeably higher then after it passed you by and is moving away. The effect is due to the fact that the observed wavelength (and therefore frequency / pitch) of sound corresponds to the distance between two points of equal phase (i.e., two sequential wavefronts). Ultimately, the Doppler effect thus originates in a change in reference frame (the same frames we encountered in section 4.3): what you hear is indeed different from what the

[1]A wavefront corresponding to the maximum extension u is sometimes called a wavecrest.

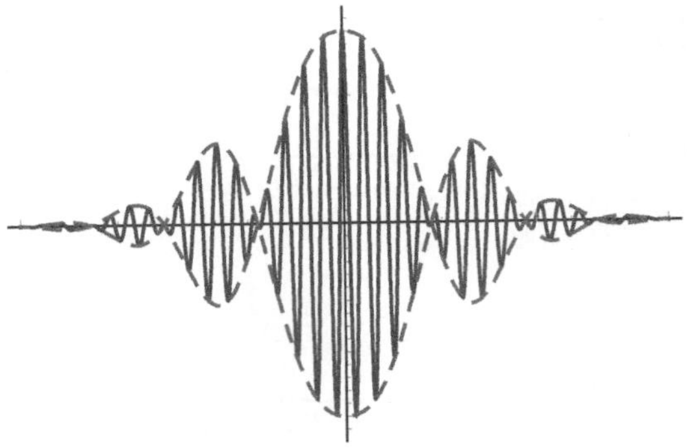

Figure 9.5: Amplitude-modulated wave. The amplitude of the carrier wave (blue, traveling at phase velocity $v_W = \omega / k$) is changed over time, resulting in an envelope (red) which travels at the lower group velocity v_g.

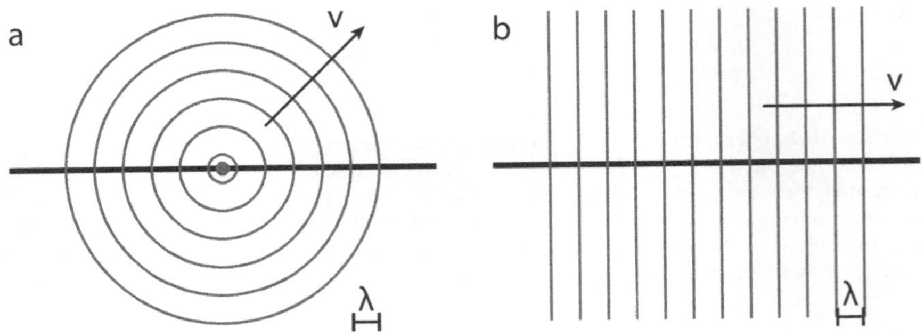

Figure 9.6: Snapshots of the wavefronts (points with equal phase) of a longitudinal wave for (a) a point source (with spherical wavefronts) and (b) a planar wave (where all wavefronts are parallel planes). Successive wavefronts are separated by one wavelength.

ambulance's driver hears. The latter is easy: the driver isn't moving with respect to the siren, so (s)he simply hears it at whatever frequency it is emitting. For the stationary observer however, the ambulance moves between emitting the first and second wave crest, and so their distance (and hence the observed wavelength / frequency) changes, see figure 9.8.

Calculating the shift in wavelength is straightforward. Let us call the speed of sound v and the speed of the source u. The time interval between two wavefronts, as emitted by the source, is Δt. In this time interval, the first wavefronts travels a distance $\Delta s = v\Delta t$, while the source travels a distance $\Delta x = u\Delta t$. For the observer to which the source is approaching, the actual distance between two emitted wavefronts is thus $\Delta x' = \Delta s - \Delta x = (v-u)\Delta t$. The actual distance between the wavefronts is the observed wavelength, λ_{obs}, while

Christian Doppler (1803-1853) was an Austrian physicist. Doppler was a professor of physics at Prague where he developed the notion that the observed frequency of a wave depends on the relative speed of the source and the observer, now known as the Doppler effect. Doppler used this principle to explain the observed colors of binary stars. The principle was developed independently by French physicist Armand Fizeau (1819-1896), and is therefore sometimes referred to as the Doppler-Fizeau effect. In 1847, Doppler moved to Selmecbánya in Hungary, but was forced to leave again soon afterwards due to political unrest in 1848, moving to the university of Vienna. During a visit to Venice in 1853, Doppler died of pulmonary disease, aged only 49.

Figure 9.7: Christian Doppler [24].

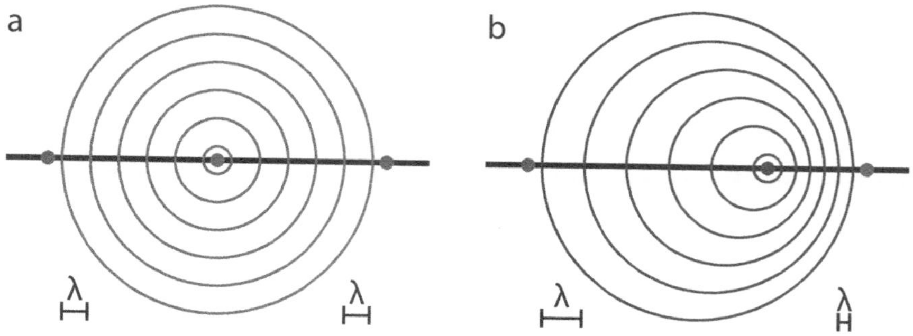

Figure 9.8: Doppler effect. The source emits waves with a fixed time interval Δt between successive wavefronts. (a) For observers that are stationary with respect to the source, the distance between wavefronts (in red) is fixed, so they measure the same wavelength as the one emitted by the source. (b) If the source is moving with respect to the observers (here to the right), the observers measure a different distance between arriving wavefronts - compressed (so shorter wavelength / higher frequency) if the source is approaching (green dot), expanded if the source is receding (purple dot).

Ernst Mach (1838-1916) was an Austrian physicist. Mach was a professor of mathematics and later physics at Graz, Prague and Vienna. His experimental work focused on the properties of waves, especially in light, as well as on the Doppler effect in both light and sound. In 1888, Mach used photography to capture the shock waves created by a supersonic bullet. In addition to physics, Mach was highly interested in philosophy, holding the position that only sensations are real. Consequently, Mach refused to accept that atoms are real, as they could not be observed directly at the time; it was Einstein's 1905 work on Brownian motion that eventually proved him wrong. The ratio of an object's speed to the speed of sound is now known as the Mach number in his honor.

Figure 9.9: Left: Mach's picture of the shockwave of a supersonic bullet [25]. Right: Ernst Mach (1902) [26].

the emitted wavelength is $\lambda = \Delta s = v\Delta t$, so the two are related through

$$\lambda_{\text{obs}} = \frac{v-u}{v}\lambda. \tag{9.19}$$

For a source that is moving away, we simply flip the sign of u; naturally for a stationary source we have $\lambda_{\text{obs}} = \lambda$. Note that we could also consider a stationary source and a moving observer: the effect would be exactly the same, where in equation (9.19) we define motion towards the source to be the positive direction.

The Doppler effect is usually expressed in terms of frequency instead of wavelength, but that is a trivial step from equation (9.19), as $f_{\text{obs}} = v/\lambda_{\text{obs}}$ and $f = v/\lambda$, which gives:

$$f_{\text{obs}} = \frac{v}{v-u}f. \tag{9.20}$$

Although we discussed the Doppler effect here in the context of sound waves, it occurs for any kind of waves - most notably also light. We will encounter it again when we discuss waves in special relativity (where speeds become comparable to that of light) in section 15.3.

Note that equation (9.19) predicts that the wavelength is zero if the speed of the source equals the speed of the emitted waves. This case is illustrated in figure 9.10a, which shows that the wavefronts pile up. For a source moving faster than the wave speed (figure 9.10b), the waves follow the source, creating a conical shock wave, with opening angle given by

$$\sin\theta = \frac{vT}{uT} = \frac{v}{u}. \tag{9.21}$$

Both the bow wave of a boat and the sonic boom of a supersonic jet are examples of shock waves.

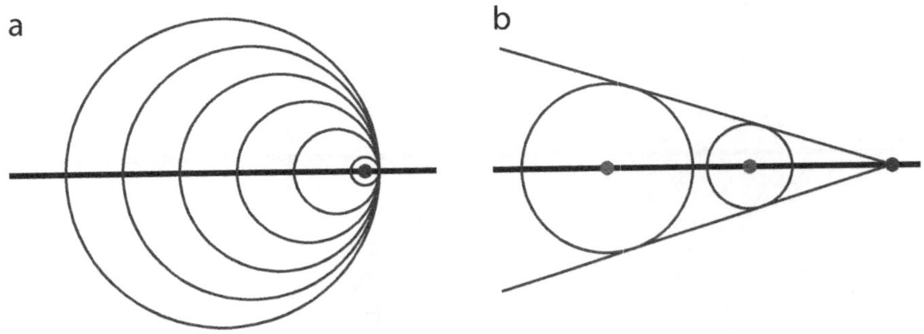

Figure 9.10: Shock waves. Like in figure 9.8, the source emits wavefronts with a fixed time interval Δt. (a) If the source is moving at the same speed as the emitted wave, the wavefronts all collide, creating a shock wave. (b) For a source moving faster than the speed of the waves, the waves all travel behind the source, creating a conical (shock) wave front (as you may have heard after seeing a jet fighter pass overhead). The blue dot indicates the current position of the source, the green dot that one period ago, the red dot two periods ago.

9.8 PROBLEMS

9.1 **Sound waves in a spring**. In section 9.2, we found that the speed of a wave in a string is given by $v = \sqrt{T/\mu}$, with T the tension in the string and μ its mass density (equation 9.7).

(a) A spring of mass m and spring constant k has an unstretched length L_0. Find an expression for the speed of transverse waves on this spring when it's been stretched to a length L.

(b) You measure the speed of transverse waves in an ideal spring under stretch. You find that at a certain length L_1 it has a value v, and at length $2L_1$ the wavespeed has value $3v$. Find an expression for the unstretched length of the spring in terms of L_1.

(c) A uniform cable hangs vertically under its own weight. Show that the speed of waves on the cable is given by $v = \sqrt{zg}$, where z is the distance from the bottom of the cable. You may assume that the stretching of the cable is small enough that its mass density can be taken to be uniform.

(d) Show that the time it takes a wave to propagate up the cable in (1c) is $t = 2\sqrt{L/g}$, with L the cable length.

9.2 In deep water, the speed of surface waves depends on their wavelength:

$$v = \sqrt{\frac{\lambda g}{2\pi}}. \tag{9.22}$$

(a) Apart from satellite images, offshore storms can also be detected by watching the waves at the beach. Equation (9.22) tells us that the longest-wavelength waves will travel the fastest, so the arrival of such waves, if their amplitude is high, is a foreboding of the possible arrival of a storm (the friction between the wind and the water being the source of the waves). A typical storm may be thus detected from a distance of 500 km, and travel at 50 km/h. Suppose the detected waves have crests 200 m apart. Estimate the time interval between the detection of these waves and the arrival of the storm (in the case the storm moves straight towards the beach).

In shallow water, the speed of surface waves becomes (to first order) independent of the wavelength, but scales with the depth of the water instead

$$v = \sqrt{gd}. \tag{9.23}$$

(b) Next to storms, a possible source of surface waves in the ocean are underwater earthquakes. While storms are typically more dangerous at sea, the waves generated by earthquakes are more dangerous on land, as they may result in tsunami's: huge wavecrests that carry a lot of energy. At open sea, the amplitude of the waves that will create the tsunami may be modest, on the order of 1 m. What will happen with this wave's speed, amplitude, and wavelength when it approaches the land?

9.3 Because the wave equation is linear, any linear combination of solutions is again a solution; this is known as the principle of *superposition*, see section 9.4. We will consider several examples of superposition in this problem. First, consider the two one-dimensional sinusoidal traveling waves $u_{\pm}(x,t) = A\sin(kx \pm \omega t)$.

(a) Which wave is traveling in which direction?

(b) Find an expression for the combined wave, $u(x,t) = u_+(x,t) + u_-(x,t)$. You may use that $\sin(\alpha) + \sin(\beta) = 2\sin((\alpha+\beta)/2)\cos((\alpha-\beta)/2)$.

(c) The combined wave is a standing wave - how can you tell?

(d) Find the positions at which $u(x,t) = 0$ for all t. These are known as the *nodes* of the standing wave.

(e) Find the positions at which $u(x,t)$ reaches its maximum value. These are known as the *antinodes* of the standing wave.

Next, consider two sinusoidal waves which have the same angular frequency ω, wave number k, and amplitude A, but they differ in phase:

$$u_1(x,t) = A\cos(kx - \omega t) \quad \text{and} \quad u_2(x,t) = A\cos(kx - \omega t + \phi).$$

(f) Show that the superposition of these two waves is also a simple harmonic (i.e., sinusoidal) wave, and determine its amplitude as a function of the phase difference ϕ.

Finally consider two sources of sound that have slightly different frequencies. If you listen to these, you'll notice that the sound increases and decreases in intensity periodically: it exhibits a beating pattern, due to interference of the two waves in time. In case the two sources can be described as emitting sound according to simple harmonics with identical amplitudes, their waves at your position can be described by $u_1(t) = A\cos(\omega_1 t)$ and $u_2(t) = A\cos(\omega_2 t)$.

(g) Find an expression for the resulting wave you're hearing.

(h) What is the frequency of the beats you're hearing? NB: because the human ear is not sensitive to the phase, only to the amplitude or intensity of the sound, you only hear the absolute value of the envelope. What effect does this have on the observed frequency?

(i) You put some water in a glass soda bottle, and put it next to a 440 Hz tuning fork. When you strike both, you hear a beat frequency of 4 Hz. After adding a little water to the soda bottle, the beat frequency has increased to 5 Hz. What are the initial and final frequencies of the bottle?

9.4 One of your friends stands in the middle of a rectangular 10.0×6.0 m swimming pool, his hands 1.0 meter apart in the direction parallel to the long edge of the pool. He produces surface waves in the water of the pool by oscillating his hands. At the edge, you find that at the point closest to your friend, the water is rough, then if you move to the side, it gets quiet, rough again, and quiet again. That point, where the water gets quiet for the second time, lies 1.0 m from your starting point (facing your friend).

(a) What is the wavelength of the surface waves in the pool?

(b) At which distance does the water get quiet for the first time?

(c) And at which distance do you find rough water for the third time (counting the initial point)?

9.5 The Doppler effect is the shift in observed frequency of a wave due to either a moving observer or moving source, as discussed in section 9.7. We will consider a sound wave emitted by some noisy source and observed by you.

(a) If you are standing still and the source is moving towards you, will the frequency you hear be higher or lower than the frequency emitted by the source?

(b) If you move towards a stationary source, will the frequency you hear be higher or lower than the frequency emitted by the source?

(c) The observed frequency f_{obs} depends on the actual frequency emitted by the source f_{source} (obviously), the speed of the source v_{source}, the speed of the observer v_{obs} and the speed of sound v_{sound}. Take the observer to be stationary. What happens if the source is stationary also? And what if the source moves at the speed of sound?

(d) Based on your answers to the previous items, guess a functional form for f_{obs} as a function of f_{source}, v_{source} and v_{sound}.

(e) Now lets do the math properly. Assume $0 < v_{\text{source}} < v_{\text{sound}}$, and $v_{\text{obs}} = 0$. Also assume the source moves in a straight line towards the observer. Plot the position of the source for a few points in time, and the fronts of the emitted waves at the *same* point in time (so not the time of emission - then they are all at the source, but the time of last emission, or slightly after that).

(f) Calculate the period T_{obs} measured by the observer by considering the arrival times of two maxima emitted a time interval T_{source} apart. *Hint*: first find the distance ΔL the observer would measure between these maxima, i.e., the observed wavelength.

(g) Convert your answer to the previous point to an expression for f_{obs}.

(h) Was your guess at (5d) right? Which aspects of the correct answer could you *not* have guessed based on your observations at (5c)?

(i) How does the expression for f_{obs} change if the observer is moving? And what if both source and observer move?

9.6 A supersonic (faster than sound) object (e.g. a fighter plane) is moving towards you at speed v_{source}, creating a shock wave in the shape of a cone. Show that the opening angle of this cone is given by $\sin\theta = v_{\text{sound}}/v_{\text{source}}$ (equation 9.21). *Hint*: draw the positions of the sound wave as emitted one, two and three periods ago (as in figure 9.10b), then use geometry to find the angle their tangent line makes with the line of flight.

II

Special relativity

10

EINSTEIN'S POSTULATES

10.1 AN OLD AND A NEW AXIOM

The theory of special relativity is built on two postulates (our axioms for this chapter). The first one also applies to classical mechanics, and simply states that:

Axiom 1 (Principle of relativity). *The laws of physics are identical in every inertial reference frame.*

You probably haven't heard of 'inertial reference frames' before. Quite likely, you've not given this principle much thought either, but nonetheless, you are (almost certainly) intimately familiar with it on an intuitive level. Consider an example we'll use a lot in this chapter. Suppose you're in a train car with no windows. Is there any experiment you can devise within the confines of the car that will tell you whether the train is standing still or moving at constant velocity (both direction and magnitude)? The answer is no, of course - that's a direct consequence of Newton's first law. A pendulum will hang straight down in a stationary train and in one moving at constant velocity, and a ball you roll will trace out a straight line in both cases. Things change of course when the train accelerates (that's where Newton's second law comes in), but as long as you keep your speed (zero or not) and direction fixed, you might as well be stationary as far as physics is concerned.

Now what's an inertial reference frame? A reference frame is simply the set of measures you use to describe the world: your coordinate system. For the person on the platform, the system will be fixed to the platform, with the origin (for example) at the point they're standing. For someone on the train, it'd be convenient to have the origin at the corner of the car, and of course the frame co-moving with the car. An *inertial reference frame* could be defined either as any reference frame that moves at constant velocity with respect to another inertial reference frame, or (as is most often done), by inverting the principle of relativity, stating that an inertial reference frame is one in which the laws of physics (i.e., Newton's laws in classical mechanics, or more specifically Newton's first law) hold without modifications. To illustrate that this is not a trivial point, consider a rotating reference frame: there the laws of physics actually change (you experience additional forces like the centrifugal and Coriolis force), and you could figure out you're rotating from a simple experiment (a pendulum at rest will no longer point down, but slightly outward).

Returning to inertial reference frames, there is one more point to be made, which you also already know. Both position and velocity are *relative* concepts, in the sense that they depend on the observer. In the train example this is obvious. From the point of view of a person sitting on the train, other objects on the train are stationary in their comoving reference frame, so at a fixed position and zero velocity; the observer on the platform however will tell you that the same object has a changing position and a velocity equal to that of the moving train. In classical mechanics, what is not relative is *acceleration.* As Newton's second law holds in both inertial reference frames, the same force gives the same acceleration according to both observers. What Einstein discovered is that although this observation still holds at relatively low velocities, it is not true at higher speeds. Instead, both observers will agree on the value of a different observable quantity: the speed of light c.

Axiom 2 (Light postulate). *The speed of light in vacuum is the same in all inertial reference frames.*

The light postulate has an important consequence: it sets the speed of light as the ultimate speed limit in the universe. Worse, you (or any other object with mass), cannot even travel at the speed of light. We'll show this

Albert Einstein (1879-1955) was a German physicist, and quite likely the most widely known scientist in the world today. Einstein studied physics in Zürich, but could not find a research position after he graduated, so he combined his research work with a position at the Swiss patent office in Bern. In 1905, his 'miracle year', Einstein published four papers, all with enormous impact on physics: one explaining Brownian motion of small particles in water, one on the (quantum-mechanical) photo-electric effect, one on special relativity, and one on the energy-matter relation (the famous $E = mc^2$). This work led to Einstein becoming a professor in 1909, and receiving the Nobel prize (for the photo-electric effect) in 1921. Einstein extended the theory of relativity to include gravity, resulting in the prediction of the bending of light by gravity (1911, confirmed 1919) and the existence of gravitational waves (1915, confirmed 2015). Einstein became a public figure in the 1920s, visiting many places around the world. When the Nazi's seized power in Germany in 1933, Einstein, who was Jewish and living in Berlin, became one of the first targets. He left Germany and gave up his citizenship, eventually moving to Princeton in the US, and advocating for the active extraction of fellow Jewish German scientists. Having been a pacifist all his life, Einstein vehemently opposed war, but also realized that Nazi Germany would not hesitate to build and use an atomic bomb, so he argued that the US should develop one also (though he was horrified when it was used against Japan). In his years in Princeton, Einstein tried to find a theory unifying gravity and quantum mechanics, but failed to do so (we still haven't succeeded); he did not like the random nature inherent in quantum mechanics and tried to prove it was incomplete (formulating the Einstein-Podolsky-Rosen paradox), which, though later proven incorrect, led to the study of quantum entanglement that is the foundation of a future quantum internet.

Figure 10.1: Official 1921 Nobel prize portrait of Albert Einstein [27].

mathematically further on, but a simple thought experiment suffices to show that this point is true. Assume the opposite: suppose you can (and do) travel at light speed with respect to a stationary observer. Next, suppose you emit a light pulse, for instance by switching a flashlight on and off. From your point of view, the pulse travels at light speed, so it speeds ahead of you quickly. However, from the stationary observer's point of view, the pulse also travels at light speed - which is the same speed you travel at, so the photon would never leave you. As the photon needs to either leave you or stay with you (but cannot do both), we arrive at a logical contradiction, and conclude that you cannot travel at the speed of light.

10.2 Consequences of Einstein's postulates

The combination of the two postulates in section 10.1 leads to a number of consequences that appear to be at odds with everyday experience. They imply that there are no such things as universal measures of time and length, nor even agreement on whether events are simultaneous or not. The reason why we don't observe these consequences all the time is that their effects are very small for objects which are moving slowly (as compared to the speed of light). Nonetheless, they do exist, and can be measured - and matter a lot in situations where speeds are high, such as in particle accelerators and cosmic radiation.

10.2.1 Loss of simultaneity

Consider the following (thought) experiment. Somebody stands in the middle of a train car with mirrors at either end. The car moves with constant speed v with respect to the platform, at which we place a second observer. The person on the train holds two laser pointers, which she presses at the same time (or perhaps a device with a single button that sends out two beams, to avoid experimenter bias). According to the person on the train, the beams reach both mirrors simultaneously, as they travel at the same speed, and cover the same distance. Now according to the person on the platform, the beams *also* travel at the same speed(!) - Einstein's light postulate tells us that all observers measure the speed of light to be the same. However, according to this stationary observer, the train also moves, and thus the light beam traveling to the front of the train has to cover a greater distance than the one going to the back of the train. Consequently, the backwards-traveling beam arrives at its mirror *before* the forwards-traveling one does. We are forced to conclude that events that are simultaneous in one inertial reference frame are not necessarily simultaneous in another.

Fortunately, there is an event that both observers agree on: the fact that the two light beams, once reflected, return to the person in the middle of the train at the same time. From the point of view of the person on the train this is obvious. For the person on the platform, a simple calculation shows that the distance that the backwards-traveling beam gains on the outbound trip equals the distance it looses on the return trip, and vice versa. Different observers don't in general agree on the simultaneity (or even order) of events happening at different points in space, but they do agree on the order of events at a given point in space - which means that relativity preserves causality (the concept that causes precede effects).

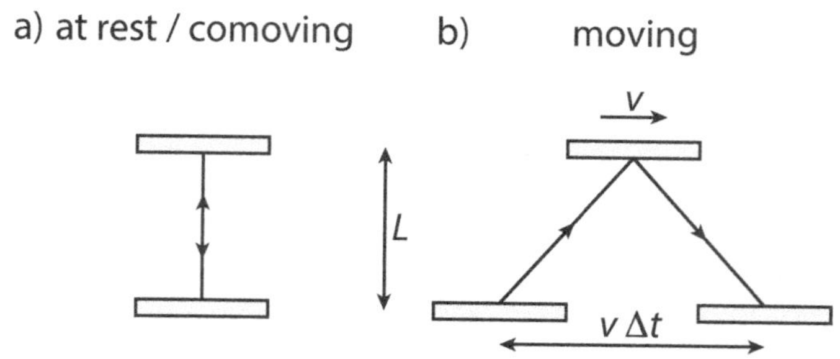

Figure 10.2: The light clock according to (a) the comoving observer and (b) a stationary observer (if the clock moves at speed v in the direction parallel to the plane of the mirrors).

10.2.2 TIME DILATION

Since everybody agrees on the speed of light (and very little else), it makes sense to use the speed of light to determine other physical quantities, such as the time interval between two events. To that end, we can construct a *light clock*: we place two mirrors in parallel, and let a beam of light bounce back and forth between them (figure 10.2). Since we agree on at least the order of events at the same point in space, let's take as our time interval the time it takes for the light to travel back and forth between the mirrors once (i.e., one round trip). According to the comoving observer (whose coordinates we'll denote with primes, and whose system of coordinates we'll call S'), this time interval is given by $\Delta t' = 2L/c$, as the distance traveled by the light is simply twice the perpendicular distance between the mirrors. However, for the stationary observer it seems that the light has moved in the direction parallel to the mirrors as well (see figure 10.2b). We'll denote his coordinate system as S, with time interval Δt. The total distance traveled by the light is (invoking the Pythagorean theorem) $\sqrt{(2L)^2 + (v\Delta t)^2}$, which should equal $c\Delta t$. Solving for Δt, we find $(\Delta t)^2 = (2L)^2/(c^2 - v^2)$, which is longer than the time interval measured by the comoving observer (that makes sense - if the light travels at the same speed, a larger distance should take longer). The two time intervals are related by:

$$\Delta t = \frac{1}{\sqrt{1 - v^2/c^2}} \Delta t' = \gamma(v)\Delta t'. \tag{10.1}$$

The factor $\gamma(v)$ will return frequently in this chapter. Its value is one if $v = 0$, and becomes progressively larger as v increases, to blow up at $v = c$. For 'small' (compared to c) values of v the value of $\gamma(v)$ is very close to one, which is why the effects of special relativity hardly ever show up in everyday life.

We have established that time intervals between two events are different for two different (comoving and stationary) observers. There is a subtle but important point to make about how time is perceived to progress according to a single observer. Let's start out with the stationary observer. Suppose this observer both measures time with a light clock in his own frame of reference, and observes an identical light clock on a moving train. Since the time interval for the roundtrip of the light in the comoving (or in this case, co-stationary) light clock is measured to be less than that of the moving clock, the stationary observer concludes that *the moving clock is running slow*. This observation is universal, and known as *time dilation*. To stress how universal time dilation is, consider the point of view of the person on the train: according to her, her clock is running 'normal', while the clock on the platform runs slow - which is in perfect agreement with the above statement, as from the point of view of the train, it's the platform that's moving.

The **Michelson-Morley** experiment (1887) was an attempt to measure the relative motion of the Earth with respect to the aether, a substance that was postulated to fill all of space. It was well known in the 19th century that the Earth's atmosphere only extends to about 100 km up, and since sunlight can reach the Earth, it was postulated that there must be another substance that acted as the medium for the propagation of light (much like sound propagates through air, water, or even solids, but not through vacuum). Since the Earth moves around the sun, it should move relatively to the aether, or from the point of view of an observer on Earth, the aether should flow through space ('aether wind'). Consequently, the aether should affect the speed of light for a beam traveling in the same direction as the wind, but not one traveling perpendicular to this direction. Michelson and Morley attempted to use this principle to measure the speed of the aether wind, with a device now known as a Michelson interferometer (pictured). However, they found no difference at all, for any angle of the interferometer arms. Lorentz (see next box) initially attempted to explain this result by introducing the concept of 'local time', which would lead to a Lorentz contraction of one of the arms with respect to the other, canceling the effect of the aether wind. Einstein took a more radical approach, dropping the concept of the aether altogether, and replacing it with his two postulates, which have Lorentz contraction as one of their consequences. Three (very sensitive) Michelson interferometers have recently been used to detect small vibrations in spacetime itself, the gravitational waves predicted by Einstein's general theory of relativity.

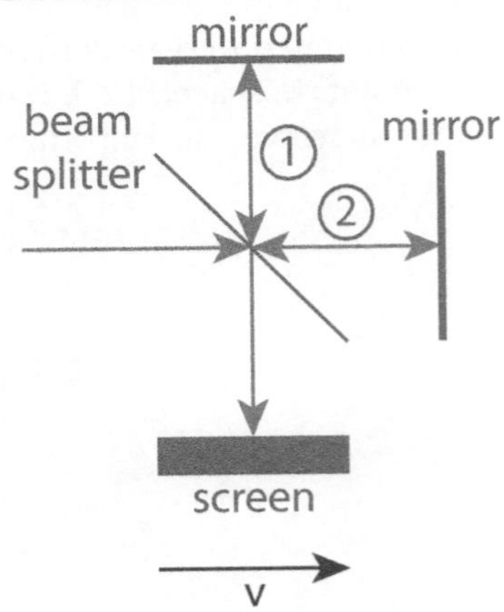

Figure 10.3: A Michelson interferometer splits a light beam in two, then measures the interference of the two beams after they have returned. If the whole system moves with speed v in the direction indicated by the blue arrow, the time it takes the light to traverse path 1 is longer than it takes to traverse path 2, resulting in interference fringes.

If the apparatus moves to the right with speed v, the speed of light on path 1 (up-down) is given by $u_1 = |\boldsymbol{u}_1| = \sqrt{c^2 - v^2}$. If the distance between the beamsplitter and the mirror is L, the time it takes to traverse path 1 (back and forth) is then given by

$$t_1 = \frac{2L}{u_1} = \frac{2L}{\sqrt{c^2 - v^2}} = \frac{2L/c}{\sqrt{1-(v/c)^2}}.$$

For path 2, the speed of the light on the way out to the mirror equals $u_{\text{out}} = c - v$, while the speed on the return path equals $u_{\text{in}} = c + v$. The total time for the trajectory (also of length L) thus equals

$$t_2 = \frac{L}{c-v} + \frac{L}{c+v} = \frac{2L/c}{1-(v/c)^2} = \frac{1}{\sqrt{1-(v/c)^2}} t_1,$$

so path 2 always takes less time than path 1. (Note that the proportionality factor is exactly the same as the one we'll find relating the time observed by a comoving and a stationary observer in the relativistic picture - Lorentz wasn't that far off!). If however the speed of light is identical in paths 1 and 2, the time it takes to traverse either of them is identical too, and we expect no interference.

A rather famous example of the effect of time dilation is the observation of the number of high-velocity muons (particles similar to electrons, but much heavier and unstable) at the surface of the earth. The muons are created in the upper atmosphere (about 20 km up), when cosmic rays collide with atmospheric atoms. The muons have a decay half-time of 2.2 μs (meaning that after this time half of the original muons have decayed). The muons are created at very high speed, close to that of light ($v = 0.999c$). Even so, classically they can only travel about 650 m before half of them are gone, and almost none will reach the surface of the earth. However, since according to us stationary observers on earth the muons' clocks run slow, we expect the half-time to be effectively extended with a factor $\gamma(0.999c) = 22$, resulting in a distance of about 15 km before half of them have decayed, and a significant number reaching the surface - as is observed.

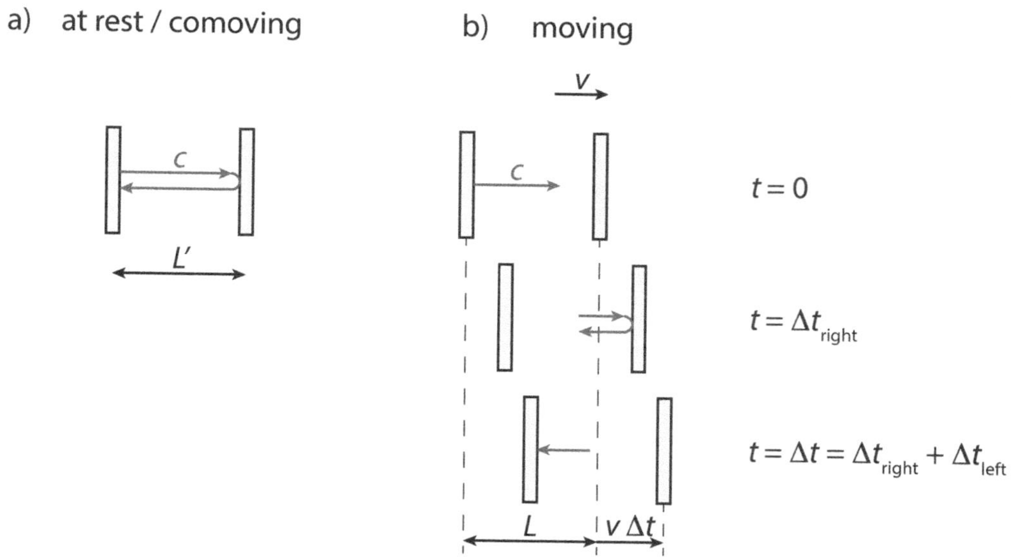

Figure 10.4: The horizontal light clock according to (a) the comoving observer and (b) a stationary observer (if the clock moves at speed v in the direction perpendicular to the plane of the mirrors).

10.2.3 LORENTZ CONTRACTION

Not only do observers in different inertial frames not agree on the duration on time intervals, they don't agree on the length of objects (i.e., the size of space intervals) either. For distances perpendicular to a direction of motion there's no issue - for the light clock in figure 10.2, both observers measure a distance L between the two mirrors. However, if they were to be asked about say the length of the wagon of the train we're imagining the moving observer to be in, their answers wouldn't agree.

To see how this difference in measured length comes about and how lengths are related, we return to the light clock, but now turn it on its side (figure 10.4)[1]. For both observers, measuring the distance between the mirrors can be done by measuring the time it takes a lightbeam to make a roundtrip between them. For the comoving observer, we find that $L' = \frac{1}{2}c\Delta t'$. For the stationary observer, the picture is more complicated, as the mirrors move while the light travels. On the way out, the distance the light has to travel is the distance L between the mirrors, plus the distance $v\Delta t_{\text{right}}$ the far mirror moves. This total distance should equal $c\Delta t_{\text{right}}$, so we get $\Delta t_{\text{right}} = L/(c-v)$. On the way back, we get a traveled distance of L minus $v\Delta t_{\text{left}}$, which should equal $c\Delta t_{\text{left}}$, so we get $\Delta t_{\text{left}} = L/(c+v)$. The total time traveled is the sum of these two, which is given by:

$$\Delta t = \Delta t_{\text{right}} + \Delta t_{\text{left}} = \frac{L}{c-v} + \frac{L}{c+v} = \frac{2Lc}{c^2 - v^2} \tag{10.2}$$

or

$$L = \frac{c^2 - v^2}{2c}\Delta t = \frac{c^2 - v^2}{2c}\gamma(v)\Delta t' = \frac{c^2 - v^2}{c^2}\gamma(v)L' = \frac{1}{\gamma(v)}L'. \tag{10.3}$$

[1] Alternatively, we may say that we measure the length of the train wagon by sending a light beam back and forth in the wagon, bouncing off a mirror at the end.

The two lengths are thus related by the inverse of the γ factor that relates two time intervals. Using an argument similar to that of the clocks (relating the length of two identical sticks, one stationary and one moving), we can conclude that *moving lengths contract*, an effect known as *Lorentz contraction*.

There is an interesting symmetry between time dilation and Lorentz contraction, which gives an alternative way of getting equation (10.3) once the effect of time dilation is known. Consider again the example of the muons, but now go to the frame co-moving with the muons. In this frame, the decay half-time is still 2.2 μs, but the same number of muons reach the surface of the earth as in the stationary frame. The reason is that according to the muons, the distance from the upper atmosphere to the surface is contracted, by exactly a factor $1/\gamma(v)$, which gives the same distance of 15 km at which half of them have decayed.

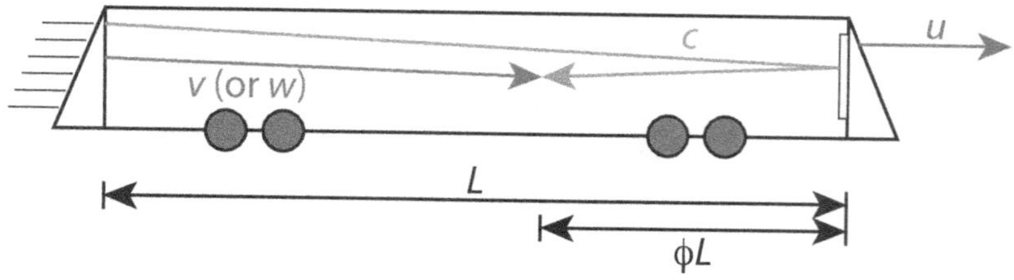

Figure 10.5: A person standing at the back of a moving train car fires a photon and a bullet at the same time. The photon reflects off a mirror at the front of the train, and meets up with the bullet sometime later.

10.3 PROBLEMS

10.1 Suppose you wake up in Rotterdam (10 km from Delft) 1 minute before class in Delft starts (i.e., at 8:44). You dress quickly, but have only 10 seconds left to get to Delft. You happen to be a very fast biker, making it exactly when your watch shows 8:45. Will the rest of the class agree that you arrived just in the nick of time?

10.2 A well-known (but probably apocryphal) Einstein quote is 'Sit on a hot stove for five minutes, and it feels like an hour. Talk to a pretty girl for an hour, and it feels like five minutes. *That's* relativity.'

(a) Einstein (at rest, frame S) sits on pins and needles for five minutes. Could there be a moving frame S' in which this same period lasts an hour? If so, determine the velocity of that frame with respect to S, if not, explain why not.

(b) Einstein talks with Marilyn Monroe for an hour. (According to another well-known anecdote, during this conversation Marilyn Monroe would have said to Einstein 'If we were to have children, and they'd have your brains and my looks, wouldn't that be fantastic?', to which Einstein replied 'Yes, but what if they'd have your brains and my looks?'). Both Einstein and Monroe are at rest in frame S. Could there be a moving frame S' in which this same period lasts five minutes? If so, determine the velocity of that frame with respect to S, if not, explain why not.

10.3 How fast would you have to fly such that you cover exactly one lightyear (as measured by a stationary observer) in one year (as measured on your clock)?

10.4 **Adding velocities** Einstein postulated that the speed of light (in vacuum, but we'll ignore that point) is the same in any inertial reference frame. Consequently, for any object with mass, the speed of light is also the absolute limit: you can never reach it, let alone exceed it. That doesn't fit well with everyday experience: if you're on a train moving at speed u, and throw a ball at speed v, an outside (stationary) observer will measure the ball's speed to be $u+v$. There seems to be no fundamental reason why we couldn't take, say, $u = v = 0.8c$, which would imply that the outside observer measures the ball to go at $1.6c$. The strange thing is that (s)he doesn't - the outside observer will tell you that they measured a speed of only $0.976c$. The reason is that although both you and the outside observer consider speed to be the distance traveled divided by the time it took to travel that distance, you no longer measure either the same distance or the same time at such high speeds. In this exercise, we'll derive a new equation for adding speeds, which shows that you can never break the speed-of-light limit.

We consider the situation sketched in figure 10.5. A train is moving with speed u. Someone standing at the back of the train car fires a photon (yellow, speed c) and a bullet (blue) at the same time. As measured on the train, the bullet has speed v; for an outside observer, the bullet has speed w. The photon is reflected by a mirror at the front of the car, and meets up with the bullet at a point a distance ϕL from the front. Note that L, the length of the car, is measured differently by the comoving and stationary (outside) observer, as is the time interval between firing and meeting of the photon and bullet. However, both observers agree that at the *events* of firing and meeting the photon and bullet are at the same place in space and time. We'll use these two points to find a relation between the speeds of the bullet as observed in the stationary and comoving frame, and the speed of the train.

(a) First, you'll take the point of view (and spacetime coordinates) of the outside observer. Denote the length of the train (as observed by you) by L, the time it takes the photon to reach the front of the train by t_1, and the time it takes to return to the meeting point by t_2. The train moves at speed u, the observed velocity of the bullet is w and that of the photon is (obviously) c. Express ct_1, the distance traveled by the photon in t_1, in terms of L, u, and t_1.

(b) Now express ct_2, the distance traveled by the photon in t_2, in terms of L, u, t_2, and ϕ.

(c) From the two equations you have, eliminate L and then rewrite the resulting equation in an expression for $t_2/t1$.

(d) Argue that $w(t_1 + t_2) = c(t_1 - t_2)$ by considering where the bullet and photon meet up.

(e) Rewrite the expression in (4d) to also give an expression for $t_2/t1$.

(f) Equate the expressions you found in (4c) and (4e), then rewrite them as an expression for ϕ.

(g) Now we'll take the comoving point of view. Find another expression for ϕ. Hint: you can either repeat the procedure above for the comoving observer, or 'translate' the expression in (4f) directly to the comoving frame.

(h) Equate the two expressions for ϕ to get an expression for w that only contains u, v and c.

(i) The rest is algebra: show that the expression you found in (4h) can be rewritten as:

$$w = \frac{v+u}{1+uv/c^2}. \tag{10.4}$$

10.5 [Challenging] Three bars are welded together into the shape of an isosceles triangle with side ratios 5:5:6. An observer looks at our triangle from a direction perpendicular to the plane it spans, as in figure a. The observer moves with relativistic velocity v in a direction parallel to the plane of the triangle.

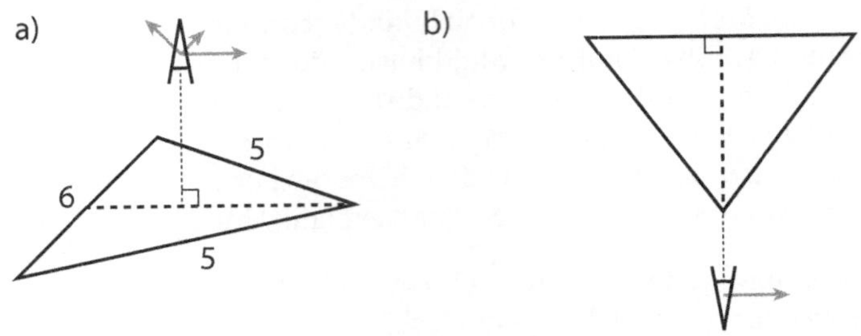

(a) In which direction and at which speed should the observer be moving so that according to him, the triangle is equilateral?

(b) We now rotate the triangle in such a way that all its corners and the observer are in one plane. The longest side of the triangle is arranged perpendicular to the observer's line of sight, and at the far end, as in figure b. The observer moves with speed v (not necessarily the same speed as in part (a) in the direction parallel to this long edge. Show that the observer sees a distorted triangle.

(c) As the observer keeps moving along the same line as in (b), show that it seems to him as if the triangle is rotating, and determine in which direction it rotates.

11

LORENTZ TRANSFORMATIONS

As we've seen at the end of section 10.1, the light postulate implies that there is a truly universal speed limit: nothing can move faster than the speed of light. This statement doesn't fit very well with the classical mechanics world you're used to. After all, if you're on a train moving at speed u, and throw a ball at speed v, a (stationary) observer on the platform will measure the ball's speed to be $u+v$. You may not be able to move at the speed of light, but there is no fundamental reason why the train could not move at, say, $u = 0.8c$, and similarly no reason why you couldn't throw the ball at $v = 0.8c$ either. However, as we'll see below, in this case the outside observer does *not* measure the ball's speed to be $1.6c$, but 'only' $0.976c$, so less than the speed of light. The reason is that the simple speed addition equation $v_{\text{outside}} = u + v$ does not hold when u or v gets close to c, another consequence of the light postulate. We'll derive the correct equation for velocity addition in this section using one method, and in problem 10.2.4 you can do it yourself using another method, based on an extension of the train example.

11.1 CLASSICAL CASE: GALILEAN TRANSFORMATIONS

To figure out how velocities add in our new reality set by the light postulate, we need to reconsider the world-view of a stationary and moving observer, each in their own inertial reference frame. In classical mechanics, for an observer moving at speed u in the x-direction, we can find the coordinates of this observer's reference frame with respect to those of a stationary observer using a simple set of transformation rules:

$$x' = x - ut, \tag{11.1a}$$
$$y' = y, \tag{11.1b}$$
$$z' = z, \tag{11.1c}$$
$$t' = t. \tag{11.1d}$$

Here the primed variables denote the coordinates of the moving observer, and the unprimed variables the stationary ones. We'll call the stationary frame S, and the moving frame S'. Of course we could also express the coordinates of S in those of S' - that's just equation (11.1) with the sign of u flipped. Note that we included the observation that time, as measured by both observers, is the same, as well as the y and z coordinates (since the train moves in the x direction - and we can just pick the x direction to be the one the train moves in). Equation (11.1) is known as the *Galilean coordinate transformation.* Note that it fits with the classical statement that accelerations are the same as measured in any reference frame:

$$a = \frac{\mathrm{d}^2 x'}{\mathrm{d}(t')^2} = \frac{\mathrm{d}^2(x-ut)}{\mathrm{d}t^2} = \frac{\mathrm{d}}{\mathrm{d}t}\left(\frac{\mathrm{d}x}{\mathrm{d}t} - u\right) = \frac{\mathrm{d}^2 x}{\mathrm{d}t^2}. \tag{11.2}$$

11.2 DERIVATION OF THE LORENTZ TRANSFORMATIONS

Instead of constant acceleration, in the theory of relativity we have a constant speed of light in each inertial reference frame. That means that the transformation rules (11.1) change. One thing that will not change is that spatial directions in which there is no motion are measured the same by all observers (our observers

Hendrik Antoon Lorentz (1853-1928) was a Dutch physicist, considered widely as the leading theoretical physicist of his time. At age 24, Lorentz became a professor at Leiden university where he initally worked on electromagnetism. He provided the theoretical explanation for the recently discovered (quantum-mechanical) Zeeman effect, for which he and Zeeman shared the Nobel prize in 1902. Around 1900, Lorentz developed the set of transformations now named after him in an attempt to interpret the results of the Michelson-Morley experiment. As Einstein built the theory of relativity on the mathematical tools provided by Lorentz, it was originally referred to as the Lorentz-Einstein theory; Lorentz himself quickly appreciated Einstein's insights and consistently referred to 'Einstein's principle of relativity'. Lorentz resigned from his position in Leiden in 1912 to have more time to do research, moving to the Teylers museum in Haarlem (still open today); Lorentz' successor in Leiden, Paul Ehrenfest, founded an institute for theoretical physics there that is now known as the Lorentz institute. From 1918 till 1926 Lorentz focussed his efforts on maritime engineering, as chair of the committee charged with designing the Afsluitdijk, a 32 km dike that closes off the former Zuiderzee in the north of the Netherlands. Lorentz solved the various necessary hydrodynamic problems numerically by hand, one of the first engineering problems approached in this way; when construction was finished, it turned out that his calculations had been highly accurate. One of the two sets of locks in the dyke is named after him.

Figure 11.1: Hendrik Antoon Lorentz, around 1916 [28].

are after all both stationary in the y and z directions), so we'll only consider the x direction, and time. To treat space and time on an equal footing, they should of course have the same dimensions. Fortunately, the one universal constant in special relativity, the speed of light c, converts a time to a space (or vice versa), so we'll consider a transformation on position x and 'time' ct. A second thing that won't change is that the transformations have to be linear. If they were not, they would violate the principle of relativity, because then the length of an object (or of a time interval) would depend on the choice of origin of our coordinate frames S and S'.

For a general linear transformation, we write:

$$\begin{pmatrix} x' \\ ct' \end{pmatrix} = A \begin{pmatrix} x \\ ct \end{pmatrix}, \qquad \text{with} \qquad A = \begin{pmatrix} a_{11} & a_{12} \\ a_{21} & a_{22} \end{pmatrix}. \tag{11.3}$$

We want our transformation to be invertible, so $\det(A) \neq 0$ and

$$\begin{pmatrix} x \\ ct \end{pmatrix} = \frac{1}{\det(A)} \begin{pmatrix} a_{22} & -a_{12} \\ -a_{21} & a_{11} \end{pmatrix} \begin{pmatrix} x' \\ ct' \end{pmatrix}. \tag{11.4}$$

We can find[1] the coefficients of A by simply demanding that S' moves relative to S at constant speed u, and the value of c is the same in S and S'. To start with the first condition, consider a stationary point in S', so $x' = b'$. From (11.3) we have $x' = a_{11}x + a_{12}ct$, or $x = (x' - a_{12}ct)/a_{11}$. On the other hand, in S this point moves at speed u, so the same point is described by an equation of the form $x = ut + x_0$. Therefore, we have $-a_{12}c/a_{11} = u$. Naturally, we can also consider a stationary point in S, $x = b = (a_{22}x' - a_{12}ct')/\det(A)$, which is moving at speed $-u$ in S', so $x' = -ut' + x'_0$, which gives $-u = a_{12}c/a_{22}$. It follows that $a_{22} = -a_{12}c/u = a_{11}$,

[1] The derivation that follows is not mathematically difficult (these are all linear equations, after all), but it contains a fairly large number of steps. The easiest way to get your head around them is to take a piece of paper and do them yourself.

and we can rewrite our transformations (11.3) and (11.4) as:

$$x' = a_{11}(x + \frac{a_{12}}{a_{11}}ct) = a_{11}(x - ut), \tag{11.5a}$$

$$ct' = a_{21}x + a_{22}ct = a_{11}(\frac{a_{21}}{a_{11}}x + ct), \tag{11.5b}$$

$$x = \frac{a_{11}}{\det(A)}(x' + ut'), \tag{11.5c}$$

$$ct = \frac{a_{11}}{\det(A)}\left(-\frac{a_{21}}{a_{11}}x' + ct'\right). \tag{11.5d}$$

Note that with $a_{11} = 1$ and $a_{21} = 0$, these are simply the Galilean transformations again. We'll allow $a_{21} \neq 0$ to accommodate for the light postulate. To see how that works, we first calculate the velocity of a moving object in either reference frame, and relate them to each other:

$$\begin{aligned} v' \equiv \frac{\mathrm{d}x'}{\mathrm{d}t'} &= \frac{a_{11}\,\mathrm{d}(x-ut)}{a_{11}\,\mathrm{d}((a_{21}/a_{11})(x/c)+t)} = \frac{\mathrm{d}x - u\,\mathrm{d}t}{(a_{21}/ca_{11})\,\mathrm{d}x + \mathrm{d}t} \\ &= \frac{\mathrm{d}x/\mathrm{d}t - u}{1 + (a_{21}/ca_{11})\,\mathrm{d}x/\mathrm{d}t} = \frac{v-u}{1+(a_{21}/a_{11})(v/c)}, \end{aligned} \tag{11.6}$$

where we used $v \equiv \mathrm{d}x/\mathrm{d}t$. Inversely, we have:

$$v = \frac{\mathrm{d}x}{\mathrm{d}t} = \frac{v' + u}{1 - (a_{21}/a_{11})(v'/c)}. \tag{11.7}$$

As the light postulate states, it doesn't matter if we measure c in S or S', we always get the same number. So for light, we should have $v' = c = v$, which we can use in equation (11.6) to determine a_{21}/a_{11}:

$$c = \frac{c-u}{1+(a_{21}/a_{11})} \quad \text{so} \quad \frac{a_{21}}{a_{11}} = -\frac{u}{c}. \tag{11.8}$$

The transforms now become:

$$x' = a_{11}(x - ut), \tag{11.9a}$$

$$ct' = a_{11}\left(ct - \frac{ux}{c}\right), \tag{11.9b}$$

$$x = \frac{1}{a_{11}}\frac{1}{1-u^2/c^2}(x' + ut'), \tag{11.9c}$$

$$ct = \frac{1}{a_{11}}\frac{1}{1-u^2/c^2}\left(ct' + \frac{ux'}{c}\right). \tag{11.9d}$$

In equations (11.9) we used $\det(A) = a_{11}^2(1 - u^2/c^2)$. We are left with one undetermined parameter, the value of a_{11}. We'll use it to make the transformation symmetric - after all, we could have started with S' as stationary and S as moving (with speed $-u$), and we should get the same transforms, except for the sign of u. Equating the prefactor in equations (11.9a) and (11.9c), we find that $a_{11} = \gamma(u)$, with $\gamma(u)$ again defined as

$$\gamma(u) = \frac{1}{\sqrt{1-(u/c)^2}}. \tag{11.10}$$

Note that $\gamma(u) = \gamma(-u)$, in accordance with the earlier notion that it doesn't matter whether you are in S watching S' move at u, or in S' watching S move at $-u$. We have now arrived at the *Lorentz transformations*:

$$x' = \gamma(u)\,(x - ut), \tag{11.11a}$$

$$ct' = \gamma(u)\left(ct - \frac{ux}{c}\right), \tag{11.11b}$$

$$x = \gamma(u)\left(x' + ut'\right), \tag{11.11c}$$

$$ct = \gamma(u)\left(ct' + \frac{ux'}{c}\right). \tag{11.11d}$$

The Lorentz transformations transform both space and time. Consequently, our two observers do not only measure space differently, as in the classical system (recall the stationary and comoving coordinates), but they also measure time differently! For small speeds, $\gamma(u)$ is (very) close to one and the effect negligible, but for high speeds it certainly is not. As we have already seen in the previous section based on the train argument, and see again below, these different time measurements lead to potentially confusing results: the two observers no longer agree on which events are simultaneous, how long a meter stick is, or how long it took to travel from one place to another.

Before going to the applications, we have a few closing remarks about the Lorentz transformations. First, we put in some effort to make the transformation symmetric between going from S to S' and vice versa. We can do more though. Since time and space now both transform, and get mingled up in the transformation, it is no longer appropriate to separate them; instead, we'll consider a combined system of four dimensions known as *spacetime*. As proper physicists, we should however not compare apples and oranges, or time and space. We already converted the time coordinate to a space coordinate by multiplying it with c. In equations (11.11a) and (11.11c) we canceled that c in front of t with a c in the denominator, but it is cleaner to put it back, so we get an even better sense of the equality of space and time in the Lorentz transformation:

$$x' = \gamma(u)\,(x - (u/c)ct)\,, \tag{11.12a}$$

$$ct' = \gamma(u)\,(ct - (u/c)x)\,. \tag{11.12b}$$

Note that in equations (11.12) the velocity u only appears as a fraction of c: we only have expressions of the form u/c, making all the coefficients of our transforms nicely dimensionless.

Second, all equations in this section are for a transformation between a stationary frame and one moving in the positive x-direction with speed u. Since we're in principle free to choose our coordinates, we can always re-label or construct our axes to match this setup. In practice, that may not always be handy though, so we could also consider movement into a different direction. Of course, moving in either the y or z direction just makes those axes swap with the x axis considered here, so we won't bother to explicitly write down those transformations. We can also write down the transformation for movement in a general direction $\boldsymbol{u}$:

$$\begin{aligned} \boldsymbol{x}' &= \boldsymbol{x} + (\gamma(u) - 1)\frac{\boldsymbol{u}\cdot\boldsymbol{x}}{\boldsymbol{u}\cdot\boldsymbol{u}}\boldsymbol{u} - \frac{\gamma(u)}{c}\boldsymbol{u}ct, \\ ct' &= \gamma(u)\left(ct - \frac{\boldsymbol{u}\cdot\boldsymbol{x}}{c}\right), \end{aligned}$$

where $u = |\boldsymbol{u}|$ is the speed of the moving frame, and $\boldsymbol{x} = (x, y, z)$.

Finally, we note that the collection of Lorentz transformations in the x direction[2] form a group under composition. If a system S' moves with respect to S with velocity u, and S'' moves with respect to S' with velocity v, then you can make a Lorentz transformation from S to S' and from S' to S'', but also from S to S'' directly[3]. As you can check for yourself (Problem 11.1), the transformation from S to S'' is indeed another Lorentz transformation. The velocity of S'' with respect to S is again not $u + v$, but $(u+v)/(1+uv/c^2)$, as given by the relativistic velocity addition equation (11.14) derived below.

11.3 Some consequences of the Lorentz transformations

11.3.1 Loss of simultaneity

If two events are simultaneous for a moving observer in S', the observer measures their time interval as $\Delta t'$. If the two events happen at the same position ($\Delta x' = 0$), the Lorentz transformations give $\Delta x = 0$, $\Delta t = 0$ as well. However, if the two events in S' are spatially separate ($\Delta x' \neq 0$), we find that for an observer in S $\Delta t = \gamma(u)(u/c^2)\Delta x'$, and therefore the two events are not simultaneous. Things even get worse: suppose two events A and B happen in S a distance Δx apart, and a time interval Δt after each other. Now if $\Delta x > (c/u)c\Delta t$, a moving observer in S' will conclude that $\Delta t' < 0$, which means that event B happens *before* event A! Fortunately, this does not violate causality, as a signal from A to B (or vice versa) will at most travel with the speed of light, which, as we will see in the next section, means that for the conditions given, A and B cannot be causally connected - i.e., you cannot reverse cause and effect, no matter how fast you run.

[2] These transformations in a given direction are sometimes also referred to as Lorentz boosts.

[3] Think, for example, of S as a stationary platform, S' as a moving train, and S'' as a toy train in the moving train.

11.3.2 TIME DILATION AND LORENTZ CONTRACTION

A stationary observer in frame S' measures the time difference between two points to be $\Delta t'$ on his/her own clock, while an observer in S will measure the time difference on that (moving) clock to be $\Delta t = \gamma(u)\Delta t'$, exactly the time dilation result we found in equation (10.1). Likewise, an observer in S' will measure the length of a stationary stick to be $\Delta L'$. For an observer in S, using a method that reaches the ends of the stick simultaneously (so $\Delta t = 0$), the length is ΔL. We have $\Delta x' = \Delta L' = \gamma(u)\Delta L$, so $\Delta L = \Delta L'/\gamma(u)$, which (unsurprisingly) is the Lorentz contraction result of equation (10.3).

11.3.3 VELOCITY ADDITION

We calculated the speed of an object v as measured in S as a function of the speed v' in S' and the speed u of S' in equation (11.7). Substituting the values of the constants we found later, we get the following equation:

$$v = \frac{u+v'}{1+uv'/c^2}. \tag{11.14}$$

Equation (11.14) thus follows directly from the light postulate - that is all we used to derive it. It mathematically shows you can never add velocities in such a way as to exceed the speed of light. Setting $u = v' = c$ gives $v = c$, and for any values $u < c$, $v' < c$, you'll always get $v < c$.

Equation (11.14) holds for motion in the same direction as the motion of the reference frame - for example, if you're on a moving train, and rolling a ball down the length of the train. However, you could also roll the ball in the transverse direction (say y if we call the direction in which the train is moving x). You might think that the observed velocity for the comoving and stationary observer is the same in that case (it is for Galilean transformations), but that's not the case. We have $v_y = \mathrm{d}y/\mathrm{d}t$, and although $\mathrm{d}y$ is invariant, $\mathrm{d}t$ is not. Calculating v_y in terms of v'_y (the speed at which the moving observer rolls the ball) is straightforward though, we simply apply the Lorentz transformation to $\mathrm{d}t$:

$$v_y = \frac{\mathrm{d}y}{\mathrm{d}t} = \frac{\mathrm{d}y'}{\gamma(u)\,\mathrm{d}(t' + \frac{u}{c^2}x')} = \frac{1}{\gamma(u)}\frac{\mathrm{d}y'/\mathrm{d}t'}{1 + \frac{u}{c^2}\mathrm{d}x'/\mathrm{d}t'} = \frac{1}{\gamma(u)}\frac{v'_y}{1+uv'_x/c^2}. \tag{11.15}$$

11.3.4 EXAMPLE APPLICATION: RELATIVISTIC HEADLIGHT EFFECT

Suppose you have a light source that radiates isotropically (i.e., with the same intensity in all directions). What happens if we put the light source on a moving train? Remarkably, according to a stationary observer, the light source is *not* isotropic anymore. To understand what happens, let us as usual call the direction in which the train moves x and its speed u. A ray of light emitted by the light source in S' has a velocity $\boldsymbol{v}'$ with magnitude c (on which both observers agree), and components (v'_x, v'_y, v'_z) (figure 11.2a). Now let's consider the ray of light that moves along the y' axis. Its velocity is given by $\boldsymbol{v}' = (0, c, 0)$. We can calculate the velocity components of this ray of light in frame S using the velocity transformation equations (11.14) and (11.15), which gives $\boldsymbol{v} = (u, c/\gamma(u), 0)$ (which of course still has magnitude c). The ray of light thus picks up a component in the positive x direction, and consequently gets a smaller component in the y direction. Figure 11.2b shows the resulting light cone in the positive x direction. Its opening angle θ can be easily calculated:

$$\sin(\theta) = \frac{v_y}{|\boldsymbol{v}|} = \frac{v_y}{c} = \gamma(u). \tag{11.16}$$

For reasons that you will probably find easy to guess, this phenomenon is known as the *relativistic headlight effect.* It is observed in the radiation emitted by electrons rotating around magnetic field lines orbiting Jupiter and the sun, as well as in particle accelerators on earth.

11.4 RAPIDITY AND REPEATED LORENTZ TRANSFORMATIONS*

As stated at the end of section 11.2, the composition of two Lorentz transformations is again a Lorentz transformation, with a velocity boost given by the 'relativistic addition' equation (11.14) (you're asked to prove this in problem 11.1). You could of course repeat this process for successive transformations, but the repeated addition of velocities quickly leads to impractical expressions. You could also investigate whether the combination of a Lorentz transformation in the x direction and one in the y direction again gives a Lorentz transformation. The answer is, in general, no: it is the combination of a Lorentz transformation and a rotation. In some sense, we can also consider Lorentz transformations themselves as 'rotations' in (4-dimensional) spacetime. We'll discuss spacetime in more detail in the next two sections. Here, we'll work out a different

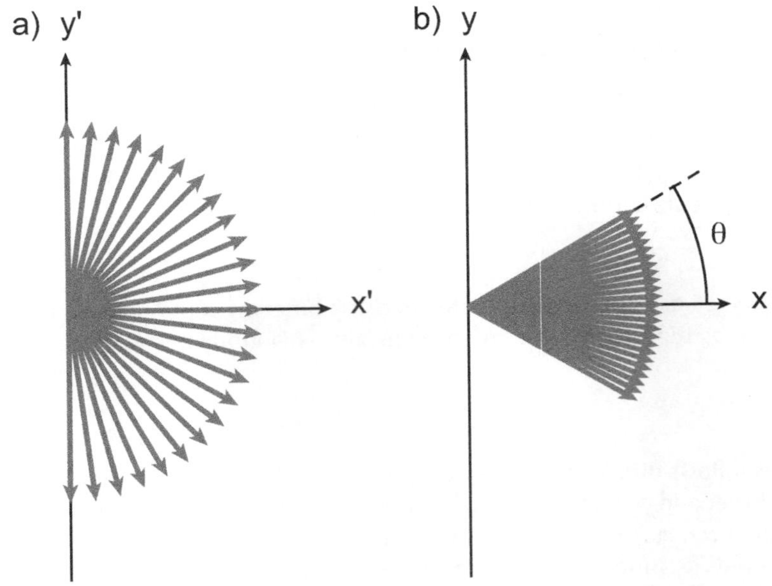

Figure 11.2: Relativistic headlight effect. (a) Isotropic light source in the comoving frame S'. (b) The same light source as observed from a stationary frame S. The opening angle θ is given by $\sin(\theta) = \gamma(u)$.

way of writing the Lorentz transformations that shows their relation to rotations. As a bonus, it will allow us to easily calculate the speed of the nth Lorentz transformation (starting from rest, all in the positive x direction).

Let us again write the Lorentz transformation as a matrix. Using the $\gamma(u)$ factor and introducing $\beta(u) = u/c$, we have

$$\begin{pmatrix} x \\ ct \end{pmatrix} = \gamma(u) \begin{pmatrix} 1 & \beta \\ \beta & 1 \end{pmatrix} \begin{pmatrix} x' \\ ct' \end{pmatrix}, \tag{11.17}$$

again beautifully illustrating the symmetry between time and space. We now define[4] the *rapidity* ϕ by

$$\tanh(\phi) = \beta(u) = \frac{u}{c}. \tag{11.18}$$

We then have

$$\gamma(u) = \frac{1}{\sqrt{1-\beta(u)^2}} = \frac{1}{\sqrt{1-\tanh^2\phi}} = \cosh\phi,$$
$$\gamma(u)\beta(u) = \frac{\beta(u)}{\sqrt{1-\beta(u)^2}} = \frac{\tanh\phi}{\sqrt{1-\tanh^2\phi}} = \sinh\phi.$$

Substituting these expressions back into the Lorentz transformations (11.17), we get

$$\begin{pmatrix} x \\ ct \end{pmatrix} = \begin{pmatrix} \cosh\phi & \sinh\phi \\ \sinh\phi & \cosh\phi \end{pmatrix} \begin{pmatrix} x' \\ ct' \end{pmatrix}, \tag{11.19}$$

which closely resembles the expression for a rotation.

We can likewise rewrite the equation for velocity addition in terms of the rapidity. Suppose we want to add velocities u and v, let the resulting velocity be w, then:

$$\tanh(\phi_w) = \frac{w}{c} = \frac{u/c + v/c}{1 + uv/c^2} = \frac{\beta_u + \beta_v}{1 + \beta_u\beta_v} = \frac{\tanh(\phi_u) + \tanh(\phi_v)}{1 + \tanh(\phi_u)\tanh(\phi_v)} = \tanh(\phi_u + \phi_v) \tag{11.20}$$

[4] If you're unfamiliar with the hypergeometric functions: they're defined like the trigonometric functions as combinations of powers of e, except that we drop the complex number i, so we have

$$\sinh(\phi) = \frac{1}{2}\left(e^{\phi} - e^{-\phi}\right), \quad \cosh(\phi) = \frac{1}{2}\left(e^{\phi} + e^{-\phi}\right), \quad \tanh(\phi) = \frac{\sinh(\phi)}{\cosh(\phi)} = \frac{e^{\phi} - e^{-\phi}}{e^{\phi} + e^{-\phi}}.$$

Note that $\mathrm{d}\sinh\phi/\mathrm{d}\phi = \cosh\phi$ like the trigonometric counterpart, but $\mathrm{d}\cosh\phi/\mathrm{d}\phi = \sinh\phi$, so no minus sign in this case.

(in problem 11.9 you get to prove the last equality). We thus find a very simple addition rule for the rapidities:

$$\phi_w = \phi_u + \phi_v. \tag{11.21}$$

Suppose now that we have a (stationary) reference system S, a system S' that moves with speed u (and rapidity ϕ) with respect to S, a system S'' that moves with speed u with respect to S', and so on. By equation (11.21), the system $S^{(n)'}$ then moves with rapidity $\phi_n = n\phi$ with respect to S. To find the relative speed u_n at which $S^{(n)'}$ moves in S, we invert the definition of the rapidity, which gives us

$$\phi = \frac{1}{2}\log\left(\frac{1+\beta}{1-\beta}\right), \tag{11.22}$$

so u_n is given by

$$u_n = \frac{1-\left(\frac{1-u/c}{1+u/c}\right)^n}{1+\left(\frac{1-u/c}{1+u/c}\right)^n}c. \tag{11.23}$$

Note that equation (11.23) provides another proof of the statement that no inertial object can move at the speed of light: for arbitrary large values of n, u_n remains less than c, so no matter how many velocity boosts you give your massive particle, you can never make it move at the speed of light, let alone exceed that speed.

11.5 PROBLEMS

11.1 Suppose system S' moves with respect to S with velocity u, and S'' moves with respect to S' with velocity v. Show that system S'' and S are related through a Lorentz transformation with velocity

$$w = \frac{u+v}{1+\frac{uv}{c^2}}.$$

11.2 Equation (11.13) gives the Lorentz transformation for an observer moving in an arbitrary direction $\boldsymbol{u}$.

(a) Show that if a ray of light is emitted in the rest frame S in an arbitrary direction, such that its trajectory is described by $\boldsymbol{x} = \boldsymbol{c}t$ (where $\boldsymbol{c} \cdot \boldsymbol{c} = c^2$), then in the moving frame S', the trajectory of the ray of light is also a straight line, given by $\boldsymbol{x}' = \boldsymbol{c}'t'$, with $\boldsymbol{c}' \cdot \boldsymbol{c}' = c^2$.

(b) What is the direction of the ray of light in S'?

11.3 (a) A male and a female student both attend lectures on relativity. Afterwards, they return home by train, moving in opposite directions, each at $(4/5)c$. Before they left, they promised to send each other messages while on the train. Unfortunately there is interference in the phone network, so they can't use their phones to do so. They do have pen and paper though, so they could write down their message and throw it to the other train. The boy, having only paid attention during the first part of the lecture (being distracted by the girl afterwards) remembers that nothing can go faster than the speed of light, and concludes that throwing something is pointless, as the (classical) relative speed of the trains exceeds the speed of light. The girl, who paid attention throughout, realizes this is not the case, and explains why in her note to the boy. At which minimum velocity should she throw the note so that it can reach the boy's train?

(b) The boy, having received the girl's message, realizes that he has a much better chance of completing the assignments if he can ask her some more questions. He therefore leaves his train, and takes the next one back (so now traveling in the same direction as the girl). Unfortunately, this train only moves at $(3/5)c$, so it won't catch up with the girl's train. The boy consoles himself with the thought that relative to him, the girl is moving, so her clock is running slow, and at least she won't have forgotten about him by the time she leaves her train. An hour passes on the boy's watch. How much time (according to him) has the girl's watch advanced in that period (assuming they both stay on their trains)?

11.4 An observer on Earth sees two spaceships (or trains, whatever you prefer) approaching from opposite directions. The observer measures their velocities in his/her own rest frame, and not knowing about relativity, uses Galilean velocity addition to conclude that the two spaceships have a relative speed of $(7/5)c$. However, an observer on one of the spaceships measures the relative speed of the other as $(35/37)c$. Find the speeds of the two spaceships relative to the Earth.

11.5 A male and a female student both attend relativity lectures. The boy however is more interested in the girl than in the lecture. Frustrated, the teacher throws a wet sponge towards him, at speed $c/2$. The girl, hoping to save the boy, tries to intercept the sponge, throwing her marker at it from the side (making a right angle, i.e., coming at the sponge from a direction perpendicular to its direction of motion, as seen in the (stationary) reference frame of the lecture hall) with speed $c/3$. For a spider who happens to sit on the sponge, what is the measured speed of the marker?

11.6 A spaceship flies away from Earth with speed $c/3$. After some time a small shuttle departs from the spaceship, in a direction that makes a right angle with that of the main ship, and a speed of $c/4$, as measured in the rest frame of the main ship. What are the magnitude and direction of the velocity of the shuttle as measured from Earth?

11.7 **Angles in Lorentz transformations**

(a) A rod moves with velocity v in a straight line relative to an inertial frame S. In its rest frame the rod makes an angle of θ' with the forward direction of its motion. Find the angle θ the rod appears to make with the direction of motion as measured in the frame S. Determine the numerical value of this angle for $\theta' = 60°$ and $v = 3c/5$.

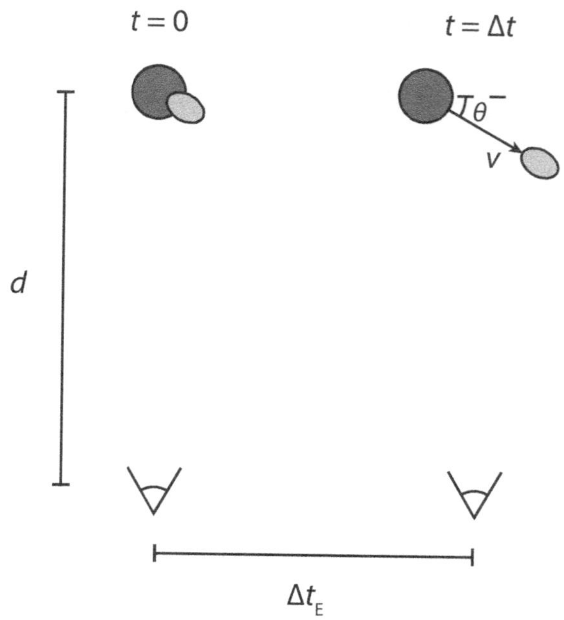

Figure 11.3: Observing the explosion of a quasar at large distance.

(b) As observed from an inertial frame S, a mirror is moving with speed v in the x direction with its plane surface perpendicular to x. Also as observed in frame S, a photon traveling in the xy-plane is incident on the mirror surface with an angle θ to the mirror's normal. Show that, as seen from S, the reflected photon has an angle $\bar{\theta}$ with the mirror normal, where

$$\cos\bar{\theta} = \frac{\cos\theta + \cos\alpha}{1 + \cos\theta\cos\alpha}, \tag{11.24}$$

and $\cos\alpha = 2(v/c)/(1+(v/c)^2)$.

11.8 Quasars are active galactic nuclei that can emit large amounts of matter, usually gas. Fortunately they are far enough away to never affect us, but close enough that we can measure e.g. the velocity of the emitted gas. These velocities occasionally seem to exceed the speed of light. To see how this can happen, consider a quasar a distance d away from Earth (as measured in the collective rest frame of the Earth and the quasar), and an explosion on the quasar resulting in such an emission at $t = 0$, see figure 11.3. The light emitted at the moment of the explosion reaches Earth at $t = d/c$. If the emitted gas is moving on a trajectory like the one shown in the figure below, light emitted from the gas has to travel a shorter distance to Earth than the light emitted at the moment of the explosion. Light emitted by an explosion on a distant quasar reaches Earth after a time interval d/c. Light emitted from the expelled gas has to travel a shorter distance. Note that the distance traveled by the gas is much smaller than the distance between Earth and the quasar, so the light arriving from both can be taken as coming from the same direction. Suppose that the gas travels at speed v.

(a) Determine the interval Δt_E between the events 'Earth observer sees the initial explosion' (which happens at the quasar at $t = 0$, but is observed later) and 'Earth observer sees the light emitted by the gas at time t''.

(b) Determine the speed of the gas an Earth astronomer would measure if they don't take the angle θ into account (we call this speed v_{obs}).

(c) Show that the observed velocity can exceed the speed of light.

(d) Show that for given actual velocity v of the gas, the observed velocity is maximized if $\sin\theta = v/c$, and that in that case we get $v_{obs} = \gamma(v)v$.

(e) What is the minimum speed v for the gas at which it can appear to have a speed equal to that of light?

11.9 [For section 11.4] Prove the last equality in equation (11.20) by expanding the hyperbolic tangents in the fraction in exponential functions.

11.10 [For section 11.4] Prove equation (11.23) by induction. If this is the first time you prove something by induction: step 1 is to prove that the equation holds for $n = 1$ (completely trivial in most cases); step 2 is to prove that if the equation holds for all values up to n, it also holds for $n + 1$. In this case, you thus have to calculate u_{n+1} by (relativistically) adding u and u_n.

12

SPACETIME DIAGRAMS

To study the effect of Lorentz transformations on the observations done by people in different inertial reference frames, we will make use of a handy tool: the spacetime diagram (figure 12.1, sometimes also called the Minkowski diagram). At first glance, these diagrams are very similar to (time, space) plots you're probably familiar with, except with the axes swapped (figure 12.1a). We plot distance horizontally, and time (or rather, c times time) vertically. A stationary object then has a vertical 'trajectory' (known as its *worldline*), and a moving object a trajectory with slope c/v, with v the object's speed. We can go to a comoving frame through a Galilean coordinate transformation (figure 12.1b, equation 11.1). The comoving frame S' has an x-axis x' that coincides with that of the stationary frame, and a time axis ct' that has the same slope as the speed of the moving object. Consequently, all point's on the object's trajectory have the same space coordinate, whereas time keeps on running at the same pace as before. (This might seem counterintuitive - didn't we change time? No, we rotated the time axis, but that doesn't change time itself, as time coordinates are determined by projecting *on* the time axis *parallel* to the space axis. Space changed: points are not projected on the same spot on the x and x' axis, even though the axes themselves coincide).

What about the Lorentz transformations? To study them, we need to consider what happens to the speed of light, which by the light postulate must remain unchanged. Because of our choice of axes (which includes a choice of units), light travels along a line of slope 1 in our spacetime diagram (figure 12.1c). Anything slower than light has a trajectory with a slope larger than 1. To have a trajectory with slope less than 1, you'd need to go faster than light. Therefore, you cannot travel from the origin into the region closer to the space than to the time axis. Objects can be there of course - but if you start at the origin, you can't travel there, or even see these objects, as not even light can travel fast enough to connect you. We'll come back to this point in section 12.2. For now, we note that the Lorentz transform must map the line of slope 1 on a line of slope 1. The only way to do so is to rotate the space and time axes both, and both by the same amount, as is done in figure 12.1d. Here time and space have both changed: the stationary (black, (x, t)) and moving (blue, (x', t')) observer measure different values for the time and space coordinates, as given by the Lorentz transformations (11.12). To find the angle α over which the axes must rotate, consider a stationary object in S': its worldline must coincide with the ct' axis, while it moves with velocity u in frame S, so the slope of the ct' axis in S must be c/u, and we have $\tan(\alpha) = u/c$.

A subtle but important point is that the Lorentz transformations do not only change the orientation of the time and space axis, but also their units (which is what you might have expected, as you know about time dilation and length contraction already). To see how this works, consider the point $(x', ct') = (0, 1)$ in S' (so a unit distance away from the origin on the time axis). The Lorentz transformations give us the coordinates of this point in S: $(x, ct) = (\gamma(u)(u/c), \gamma(u)) = \gamma(u)((u/c), 1)$. The length of this interval in S' is 1; the length of the same interval in S is $\gamma(u)\sqrt{1 + (u/c)^2}$. We conclude that the units of ct' and ct are related through this factor. A completely analogous calculation shows that the same holds for the units of x' and x (as again we might have guessed, due to the symmetry of time and space now present in our system), so we have:

$$\frac{ct' \text{ unit}}{ct \text{ unit}} = \sqrt{\frac{1 + (u/c)^2}{1 - (u/c)^2}} = \frac{x' \text{ unit}}{x \text{ unit}} \tag{12.1}$$

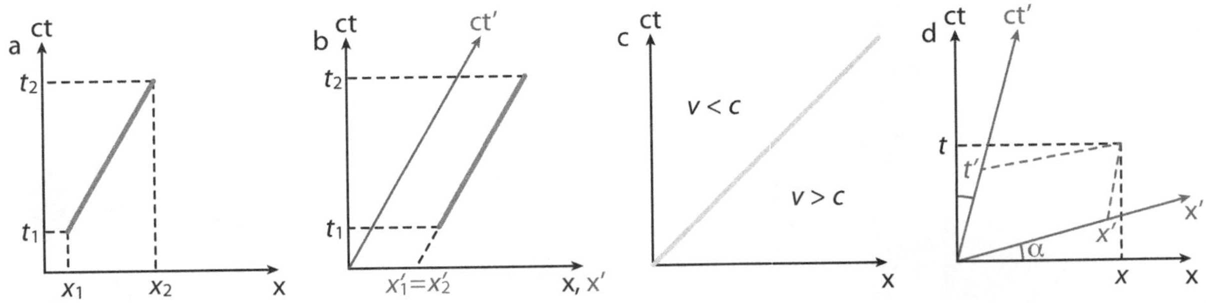

Figure 12.1: Spacetime or Minkowski diagrams. Horizontal axis depicts a spatial coordinate, vertical axis time (times c). Note that this choice of axes is opposite to the ordinary choice in classical mechanics. (a) Object (orange line) moving in space and time at speed $v = (x_2 - x_1)/(t_2 - t_1)$. A stationary object would trace out a vertical line. Time and space coordinates of a certain point are determined by projection on their respective axes, along a (dashed) line parallel to the other axis. (b) Galilean coordinate transformation to the comoving frame (x', ct'). Note that it is the time axis that turns - consequently, both the start and end points of the trajectory of the moving object are mapped to the same coordinate in x' (so the object is stationary in the comoving frame). Because the x-axis has not changed, the time coordinates in the comoving frame are the same as in the stationary one. (c) Trajectory of a light beam in a stationary frame. The light beam has speed c, which means slope 1 because we multiplied t by c. Any massive object moving in spacetime has a velocity less than c and so a slope of more than 1, and lies closer to the time axis. To have a trajectory with slope less than 1, the object would have to move faster than light. Note that objects can be present in this part of the diagram, but can't have traveled there from the origin. (d) Lorentz transformation to the frame (x', ct'). To keep the speed of light constant, the axes have to rotate by the same angle, illustrating the symmetry between space and time. Under a Lorentz transformation, both space and time coordinates of any point that is not the origin change, as illustrated - coordinates are still determined by projection on the respective axis, parallel to the other axis.

12.1 TIME DILATION AND SPACE CONTRACTION REVISITED

As we've already seen twice, Lorentz transformations do funny things with the measurement of time and space. Here, we'll study these effects once more using spacetime diagrams. First we'll consider time. Suppose you and a friend come together at some point in space and time, which we'll call O (for origin, obviously). You synchronize your (perfect) watches. Your friend next takes off at speed u in the x-direction, while you remain stationary. Your inertial frame is thus S (the black frame in figure 12.2a), whereas your friend's inertial frame is S' (blue frame in figure 12.2). After some time t_1 has passed on your watch, you whip out a telescope and take a look at your friend's watch[1] - and see that it is lagging behind yours.

To see why your friend's watch appears to be slow, consider the observation lines in figure 12.2a. You both started at the origin. You have not moved in space since, only in time, so your position in spacetime coincides with the ct axis. Your friend is moving in both time and space, so in your coordinates your friend's trajectory is a sloped line, but of course in the comoving coordinates of S' your friend is also stationary, so (s)he is moving on the ct' axis. When you look at your friend at $t = t_1$, you observe him/her at $t' = t_1'$ in S', as shown by the black arrow. From your point of view, t_1 and t_1' are *simultaneous*, because they lie on a line parallel to your space axis. However, from your friend's point of view, these events are not simultaneous at all! Instead, in S', t_1' is simultaneous with events on the line parallel to the x' axis - so point t_0 (which you already passed) on your time axis. You can find t_0 by projecting from t_1' along the x' axis on the ct axis - the green arrow in figure 12.2a. By moving away from you, your friend's time thus seems to move slower, a phenomenon we call *time dilation.* Of course, we can also find out at which point your friend's watch indicates t_1, by backtracking the upper green arrow in figure 12.2a, which shows that that time corresponds to your later time t_2.

We now have two ways to calculate how much time seems to slow down for your moving friend. On the one hand, the spacetime diagrams in figure 12.2 simply represent Lorentz transformations. Since you haven't moved in space, x=0, and equation 11.12 gives $ct' = \gamma(u)ct$, so $\gamma(u)$ is the dilation factor. Alternatively, we could use that in S, t_1 and t_1' are observed to be simultaneous. We already know that the slope of the ct' line in S is given by $\tan(\alpha) = u/c$, where α is the angle between the ct and ct' axes. We also know how the units on

[1]Well, not really - it takes time for the light from your friends watch to reach you. We'll assume you're smart, and have corrected for this already.

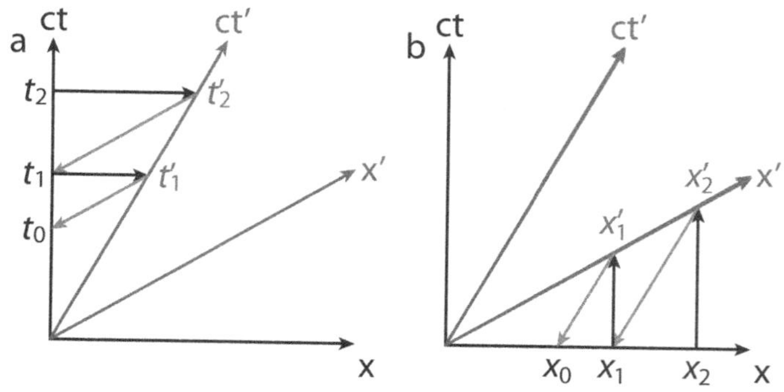

Figure 12.2: Spacetime diagrams showing (a) time dilation and (b) length contraction of a moving object (blue frame, S') as observed from a stationary point of view (black frame, S).

the ct' axis relate to those in S (equation 12.1). Putting all these together, we have:

$$\begin{aligned} t_1 &= \cos(\alpha)\cdot(\text{unit conversion factor})\cdot t'_1 \\ &= \frac{1}{\sqrt{1+(u/c)^2}}\cdot\sqrt{\frac{1+(u/c)^2}{1-(u/c)^2}}\cdot t'_1 \\ &= \gamma(u)t'_1 \end{aligned} \tag{12.2}$$

A similar thing happens for lengths. Of course, in our new four-dimensional spacetime, we would expect so, as time and space are now intimately linked. Suppose you have an object of length x_1, that you position on your x-axis at $t = 0$, with one end at the origin and one end at $x = x_1$ (figure 12.2b). Now consider an identical moving object that passes you by at speed u at $t = 0$. How long do you measure that object to be? One end is at the origin of the S' frame, which coincides with the origin of your frame; the other end sits at $x' = x'_1$ on the x'-axis, which you find by looking along your own time axis (black vertical arrow). To project back to your axis, you have to project from x'_1 on your space axis along the time axis of S' - that's the green arrow, which projects x'_1 on some point $x_0 < x_1$. From equation (11.12) we find $x'_1 = \gamma(u)x_1$, so the length of the object appears shortened by a factor $\gamma(u)$, or *Lorentz contracted.*

Note that time dilation and length contraction are really the same thing - parts a and b of figure 12.2 are identical, only reflected in the diagonal line with slope 1 that represents the path taken by light. It is therefore also not surprising that the contraction/dilation factor $\gamma(u)$ is the same in both cases.

A subtle but important point is that time dilation and length contraction are observations *you* make on an object moving *relative to you.* As long as you're in an inertial reference frame, you can always define your own frame to be the stationary one - simply pick the one that is moving with you. However, your friend can of course do exactly the same. You see your friend moving along your positive x-axis at speed u - and your friend sees you moving along their positive x-axis at speed u, the only difference being that they define 'positive' the opposite direction you do (but as this is entirely arbitrary, there is no 'right' or 'wrong' choice). Therefore, *your friend also sees your watch go slower, and your lengths contracted!*

12.2 AN INVARIANT MEASURE OF LENGTH

As we've seen in the previous section, measures of length and time intervals become relative in the new reality that is created by the light postulate. That's not very convenient: we'd like to be able to agree on the value of quantities independent of which inertial reference frame we use to measure them in. In ordinary space, the method we use to write down vectors presents us with a similar problem: we write our vector components with respect to a basis or coordinate system, and so choosing a different set of coordinates changes the numerical value of the components of the vector. Of course, the actual vector doesn't change - a physical quantity like a velocity or a force really doesn't care what specific coordinates you use to measure it. However, if you and your collaborator use different coordinates, you might run into misunderstandings when comparing notes. Fortunately, vectors have properties whose value are independent of the coordinate system used to determine their components. The most important one is their *magnitude,* which we sometimes also refer to as their *length.* If you know a vector's components, calculating its magnitude is easy: its simply the square

Hermann Minkowski (1864-1909) was a German mathematician, who used geometrical methods to address problems both in number theory and in (mathematical) physics. Born in Russia, Minkowski's Jewish family emigrated to Prussia (in present-day Germany) to avoid persecution. He worked at various universities in the German-speaking world, including ETH in Zürich, where he was one of Einstein's teachers. In 1907, Minkowski realized that the theory of relativity as introduced by Einstein two years earlier could most easily be understood in a four-dimensional spacetime in which lengths can be calculated with equation (12.4). He also developed the graphical representation named after him (figure 12.1). In September 1908, Minkowski gave a lecture titled 'Space and time' in which he laid out these ideas, starting with a now famous quote: "The views of space and time which I wish to lay before you have sprung from the soil of experimental physics, and therein lies their strength. They are radical. Henceforth space by itself, and time by itself, are doomed to fade away into mere shadows, and only a kind of union of the two will preserve an independent reality." Einstein built on Minkowski's geometrical representation to extend the special theory of relativity to the general one in the 1910's. Sadly, Minkowski did not live to see this, as he died of appendicitis at age 44.

Figure 12.3: Hermann Minkowski in 1909 [29].

root of the dot product with itself:

$$x = |\boldsymbol{x}| = \sqrt{\boldsymbol{x} \cdot \boldsymbol{x}}. \tag{12.3}$$

Similarly, although measures of both length and time change between inertial reference frame in relativity theory, there is a combination of the two whose value is independent of which reference frame you measure it in:

$$(\Delta s)^2 = (c\Delta t)^2 - (\Delta \boldsymbol{x}) \cdot (\Delta \boldsymbol{x}) = (c\Delta t)^2 - (\Delta x)^2 - (\Delta y)^2 - (\Delta z)^2. \tag{12.4}$$

Again, we see that time and space are mixed! Checking that the 'length' Δs is indeed invariant when you go from one inertial frame to another (i.e., when you apply a Lorentz transformation) is a straightforward exercise. You may wonder where the idea for equation (12.4) came from, and the answer is the same as always in relativity: from the light postulate. What both you and your collaborator will agree on is the speed of light, which (as it is a constant), you can measure in your own coordinate system as $c = (\Delta r)/(\Delta t)$, where $\Delta r = \sqrt{(\Delta x)^2 - (\Delta y)^2 - (\Delta z)^2}$ is the distance covered by a light beam, and Δt the time it took. Rewriting, you get $0 = (c\Delta t)^2 - (\Delta x)^2 - (\Delta y)^2 - (\Delta z)^2$, which won't depend on the coordinate system you used - so your collaborator gets the same answer[2].

In contrast to the length of an ordinary vector, the length of the invariant interval $(\Delta s)^2$ can become zero or even negative! As we've seen above, the interval is always zero for the path taken by a light beam (or anything traveling at the speed of light). If you travel at speeds slower than light (including standing still), the length of your interval will come out positive. A negative length would correspond to a trajectory that has a speed faster than that of light, which, as we've seen in section 10.1, is impossible. There can therefore be no communication between any two points in spacetime that are separated by an interval with negative length. In other words, they cannot be *causally connected*: it is impossible for an event at one such location to influence the other[3]. We call events separated by an interval with negative length *spacelike* connected; those connected by a positive interval are *timelike* connected (and can be causally linked), and if the interval is zero, the connection is called *lightlike*. Lightlike events can still be causally linked, as information can be send from one to another using a light signal. The timelike, lightlike, and spacelike regions correspond to the $v < c$, $v = c$ and $v > c$ regions in figure 12.1c. Together, the lightlike and timelike regions make up the *light cone*.

[2] In fact, there is an alternative derivation of the Lorentz transformations, that starts from the observation that due to the light postulate, we must have $(\Delta s)^2 = 0$ for a light beam. Additionally, space must be homogeneous (the same at every point) and isotropic (looking the same in every direction). Together with the statement that the transformations have to be linear due to the principle of relativity, you can determine the components of the transformation matrix from these observations as well.

[3] Note that we are talking about events in spacetime here: at a specific point in space *and* time. You can of course send out a signal to a specific location, but if that point is connected to the present you through a spacelike interval, you cannot influence it (or be influenced by it) at that moment, you can only influence its future. A teacher lecturing a class can thus only influence the future students, and the students in turn can only influence the future teacher, which is perhaps one reason why special relativity sometimes leads to confusion.

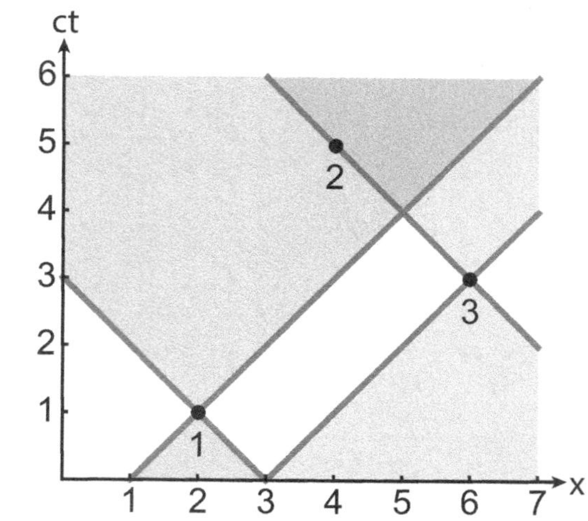

Figure 12.4: Spacetime diagrams showing the three events of the worked example in section 12.2. Events 1, 2 and 3 are indicated with black dots; the light cone of point 1 is shown in blue, that of point 3 in red. We can immediately read off that point 2 is inside the light cone of point 1 (so 1 & 2 are timelike connected), point 2 is on the edge of the light cone of point 3 (so 2 & 3 are lightlike connected) and point 3 is outside the lightcone of point 1 (so 1 & 3 are spacelike connected).

12.2.1 WORKED EXAMPLE: CAUSAL CONNECTIONS

Consider three events, occurring at spacetime coordinates (1) $(ct,x) = (1,2)$, (2) $(ct,x) = (5,4)$, and (3) $(ct,x) = (3,6)$. Which of these events can be causally connected?

SOLUTION

Method 1: We calculate the length of the invariant interval (12.4) for all three pairs:

- (1 & 2): $(\Delta s)^2 = (c\Delta t)^2 - (\Delta x)^2 = (1-5)^2 - (2-4)^2 = 12 > 0$ so this interval is timelike.
- (1 & 3): $(\Delta s)^2 = (c\Delta t)^2 - (\Delta x)^2 = (1-3)^2 - (2-6)^2 = -12 < 0$ so this interval is spacelike.
- (2 & 3): $(\Delta s)^2 = (c\Delta t)^2 - (\Delta x)^2 = (5-3)^2 - (4-6)^2 = 0$ so this interval is lightlike.

We conclude that events (1 & 2) and events (2 & 3) can be causally connected, but events (1 & 3) cannot. *Method 2*: We draw a spacetime diagram with the three events, and their (past and future) light cones, see figure 12.4. We can immediately read off that events (1 & 2) and events (2 & 3) can be causally connected (as they are in or on each other's light cone), but events (1 & 3) cannot, as they are outside each other's light cone.

12.2.2 THE INVARIANT INTERVAL AND THE ORDERING OF EVENTS

We've already seen in section 10.2 that moving and stationary observers will only agree on whether two things happen simultaneously if they happen at the same point in space and time. Suppose now that we have two events that according to a stationary observer happen in a certain order. Would a moving observer at least agree to that ordering? The answer is, in general, no: if the two events are separated by a spacelike interval, there is a finite speed (less than c) at which an observer has to move for the events to occur at the same time on their clock - and if they move faster, the ordering is reversed! Calculating this speed is straightforward: if the separation intervals in the stationary observer's frame are $c\Delta t$ and Δx, we simply find the velocity u for which the Lorentz transformed time equals $c\Delta t' = \gamma(u)(c\Delta t - (u/c)\Delta x) = 0$, so

$$u = (c\Delta t/\Delta x)c. \tag{12.5}$$

If the separation between the two events is lightlike, equation (12.5) tells us that we'd have to move at the speed of light for them to occur at equal times at our clock. For timelike separations, we'd need to move even faster than light - so then we're forced to agree on the temporal order of events. However, once again space and time show their dual nature: for a timelike separation, there may not be a speed at which the events

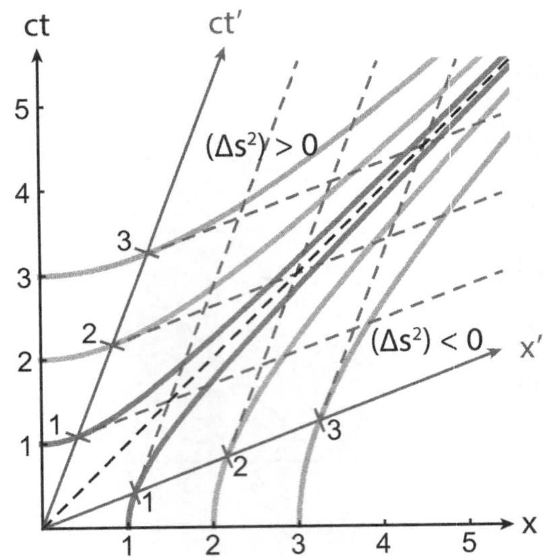

Figure 12.5: Spacetime diagrams showing hyperbolas of invariant $(\Delta s)^2$ in both the stationary (black frame, S) and moving (blue frame, S') coordinates. By finding the points where a hyperbola intersects the time or space axis of the moving frame, we can graphically construct the units in that frame, given the units in the stationary frame. Note that for any value of $(\Delta s)^2$, the hyperbola has the light cone, with $(\Delta s)^2 = 0$, as its asymptote.

occur at the same time for a moving observer, but there is one for which they occur at the same location. Unsurprisingly, that speed is simply $\Delta x/\Delta t$ - this specific result holds in classical mechanics as well[4].

12.2.3 UNITS IN SPACETIME DIAGRAMS REVISITED

We can use the invariance of the length of the interval $(\Delta s)^2$ to graphically construct the units of time and space in a spacetime diagram that has undergone a Lorentz transformation, see figure 12.5. For fixed value of $(\Delta s)^2$, the line given by equation (12.4) becomes a hyperbola in a Minkowski diagram. For $(\Delta s)^2 = 0$, we get the straight line with a 45° angle describing the path of a ray of light (dashed black line in figure 12.5). For positive values of $(\Delta s)^2$, we get hyperbolas that intersect the ct axis at $c\Delta t = \Delta s$, while asymptotically approaching the $(\Delta s)^2 = 0$ line for large values of Δx. These lines thus also intersect the ct' axis, at which point $c\Delta t' = \Delta s$. Likewise, for negative values of $(\Delta s)^2$, we can map points on the x axis to points on the x' axis.

12.3 WORLDLINES AND PROPER TIME

The trajectory followed by a particle through space and time is commonly called a *worldline* in the theory of relativity. In general, a worldline does not have to be a straight line in a spacetime diagram - you can of course speed up and slow down (relative to a stationary observer) as you please. Any clock you bring with you will, from your point of view, keep on ticking at the same rate as always. For the stationary observer its rate will however depend on your speed, and if you change direction or accelerate, that will have an effect as well. We call the time recorded by a clock in the comoving reference frame of a particle the particle's *proper time*, usually denoted by τ. We can calculate the proper time with respect to a stationary observer by chopping up the particle's trajectory into small pieces. In each of the pieces, the particle's velocity will be constant, and we can calculate the time dilation as we did before, then we'll integrate over the pieces to get the total proper time. To see how this is done, consider a part of the trajectory between $t = t_i$ and $t = t_{i+1}$, in which the particle travels from x_i to x_{i+1} as measured by a stationary observer. That observer then calculates the particles speed in this interval to be

$$v_i = \frac{x_{i+1} - x_i}{t_{i+1} - t_i}. \tag{12.6}$$

[4]As you can verify next time you're on a train: from your stationary point of view, the platforms you encounter will all pass through the same point next to you.

We now define the interval of proper time as the length of the invariant interval Δs_i corresponding to the displacement divided by the speed of light, which gives:

$$\Delta\tau_i = \frac{\Delta s_i}{c} = \frac{1}{c}\sqrt{c^2(t_{i+1}-t_i)^2-(x_{i+1}-x_i)^2}$$
$$= (t_{i+1}-t_i)\sqrt{1-\frac{(x_{i+1}-x_i)^2}{c^2(t_{i+1}-t_i)^2}} = \Delta t_i\sqrt{1-\frac{v_i^2}{c^2}} = \frac{\Delta t_i}{\gamma(v_i)}. \quad (12.7)$$

$\Delta\tau_i$ is the time interval as measured on a comoving clock. Unsurprisingly, it is related to Δt_i through the time dilation factor $\gamma(v_i)$. To calculate the total time which has passed on the comoving clock, we simply sum over all discrete intervals i. In the limit where the length of the interval becomes infinitesimally short, that sum becomes an integral, and we can calculate the proper time by:

$$\Delta\tau = \int_{t_a}^{t_b}\sqrt{1-\frac{v(t)^2}{c^2}}\,dt, \quad (12.8)$$

where $[t_a, t_b]$ is the time interval as measured by the stationary observer. Note that equation (12.8) holds for any kind of motion, not just motion at constant velocity.

12.4 PROBLEMS

12.1 In the (ct, x) coordinates of an inertial frame S three events occur at $G_1 = (1,2)$, $G_2 = (5,4)$ and $G_3 = (3,6)$. Which of these events could be causally linked?

12.2 Two astronauts (we'll call them A and B) leave Earth on January 1st in opposite directions. Astronaut A initially travels as a speed $4c/5$, and B with $3c/5$. The astronauts keep traveling until they measure their distance to Earth at 5 light years, at which point they turn around and travel back. On the way home, astronaut A travels at $3c/5$, while B travels at $4c/5$.

(a) Determine how far each astronaut has traveled in Earth's frame of reference before (s)he returns.

(b) Draw a spacetime diagram showing the trips of the two astronauts. NB: You may want to take your space for this, it will help see details clearly.

(c) Which astronaut turns around first?

(d) Find the length of the spacetime interval between the events 'astronaut A reaches his turning point' and 'astronaut B reaches her turning point', as measured by each of the astronauts, and by an observer back on Earth, and verify that the spacetime interval is indeed invariant.

(e) Which astronaut returns to Earth first?

(f) As measured by astronaut A, what is the total distance that astronaut B travels?

(g) Before they left Earth, the astronauts promised to send each other a message by radio every new year. How many such messages does each astronaut send?

(h) Draw the lines representing the radio signals of each astronaut in the spacetime diagram.

(i) For each astronaut, determine how many signals they sent on the way out, way back, and when already back on Earth.

(j) After both astronauts have returned to Earth, which of them has aged more?

12.3 You observe a spaceship with astronauts A and B (fill in your own favorite SciFi characters if you like) in it, moving at constant velocity $u > 0$ along the x-axis of your rest frame S. At time $t = t_0$, when the spaceship is at (x_0, y_0, z_0), astronaut B gets on board a small shuttle and leaves with constant velocity $v > u$ in the x-direction to a distant star, which is stationary in your rest frame S. At $t = t_0$, you measure the distance between the spaceship and distant star to be d. Astronaut B reaches the star at $t = t_1$ on your clock, does whatever (s)he wanted to do quickly, then returns to the mother ship with the same speed v, reaching it at $t = t_2$ on your clock. The mother ship meanwhile kept the same speed u throughout.

(a) Draw a spacetime diagram with the worldlines of astronauts A and B. Indicate all relevant quantities, and specify the slopes of any lines you draw.

(b) Express t_2 and x_2 (the place where the shuttle and mother ship meet) in terms of the quantities x_0, t_0, u, v, and d.

(c) Both astronauts carry high-precision watches which they synchronize when B leaves the mother ship. Express the times given by both watches upon return of B in terms of the given quantities.

(d) Which astronaut has aged more between the departure and return of B?

12.4 Having negotiated a cease-fire, an imperial and a rebel spaceship leave their meeting place with synchronized clocks (we'll set the meeting place at $x = 0$, and the time of leaving at $t = 0$), departing with relative velocity (i.e, the velocity at which they each measure the other ship to be moving) of $(3/5)c$ in opposite directions. Once he has thought about it for a day, the imperial captain comes to the conclusion that the emperor will reward him higher for eliminating the rebels than making peace with them, so he fires a photon torpedo at the rebel ship. Likewise, having pondered the option of peace for a day, the rebel captain decides that offense is the best defense, and fires a photon torpedo at the imperial ship. Both ships are equipped with launchers that can fire such torpedos with speed $(4/5)c$ with respect to the ship.

(a) Draw a spacetime diagram with the two spaceships and the two photon torpedos as seen from the imperial ship. Use units of $c\cdot$(half a day) on the time axis, and half a light-day on the space axis. Provide the calculation for each value that you have to compute. NB: make your drawing large enough for details to be properly visible and use a ruler; don't worry about the paper, we promise to recycle it in due time.

(b) When, as measured on the clock of the imperial ship, does the torpedo fired by them hit the rebel ship?

(c) When, as measured on the clock of the imperial ship, does the torpedo fired by the rebels hit them?

(d) Fortunately for the rebel ship, their shield is quite strong, being able to withstand an impact with a momentum up to $3.0 \cdot 10^{11}$ kg·m/s. The torpedo fired by the imperial ship has a mass of 1000 kg. Does the rebel ship survive the impact?

(e) Both ships have a home base 1 light week away from the meeting point. When the rebels at the home base see the imperial ship fire on their friends, they decide to mount a counterattack. They launch another ship at $v = (3/5)c$, which immediately after launch also fires a torpedo with $(4/5)c$ relative to the ship. Does that torpedo hit the imperial ship before it returns home? To answer this question, you may assume that the imperial and rebel ship left the meeting point with equal speeds, as measured from the rest frame of the bases. You may also find drawing another spacetime diagram useful.

12.5 A train with proper length L moves at speed $4c/5$ with respect to the ground. A passenger lets a toy train run from the back of the (moving) train to the front, with speed $c/3$ as measured on the large train.

(a) Draw a spacetime diagram with the worldlines of the front of the large train, the back of the large train, and the small train, from the point of view of a (stationary) observer on the ground.

How much time does it take the toy train to cross the train, and what distance does the toy train cover, in:

(b) The frame of the large train?

(c) The frame of the toy train?

(d) The frame of a stationary observer on the ground?

(e) Verify that the length of the invariant interval (equation 12.4) is the same in all three frames.

13

POSITION, ENERGY AND MOMENTUM IN SPECIAL RELATIVITY

13.1 THE POSITION FOUR-VECTOR

As we've seen in the previous section, we can define a 'length' Δs that is invariant under Lorentz transformations (equation 12.4):

$$(\Delta s)^2 = (c\Delta t)^2 - (\Delta \boldsymbol{x}) \cdot (\Delta \boldsymbol{x}) = (c\Delta t)^2 - (\Delta x)^2 - (\Delta y)^2 - (\Delta z)^2.$$

As is clear from the definition of Δs, to get invariant quantities, we should not think of space as measured in three dimensions, but of *spacetime*, measured in four dimensions. This four-dimensional world of special relativity is called *Minkowski space*, and its vectors have four components: one for time and three for space. Conventionally, we add the time component as the zeroth component of the vector. To distinguish between 'ordinary', three-dimensional vectors (which are represented in bold) and four-vectors, we'll put a line on top of the latter. The *position four-vector* is then given by:

$$\bar{\boldsymbol{x}} = (x_0, x_1, x_2, x_3) = (ct, x, y, z). \tag{13.1}$$

We would like to be able to determine the length of the position four-vector by taking the inner product of the vector with itself. However, the regular inner product is not going to work, because instead of $(ct)^2 + x^2 + y^2 + z^2$, the quantity that is independent of the reference frame is $(ct)^2 - x^2 - y^2 - z^2$. We therefore define the inner product of two four-vectors $\bar{\boldsymbol{a}}$ and $\bar{\boldsymbol{b}}$ as

$$\bar{\boldsymbol{a}} \cdot \bar{\boldsymbol{b}} = a_0 b_0 - a_1 b_1 - a_2 b_2 - a_3 b_3. \tag{13.2}$$

As we've already seen, the magnitude of the position four-vectors, as determined by its inner product with itself (equation (12.4)) is independent of the inertial reference frame you use to measure it in. In problem 13.3 you'll show that consequently, the value of the inner product of any two four-vectors is reference frame independent.

It may seem that we're back to normal - we've added a dimension, and introduced a new inner product, but with those, we should be able to do calculations just as easily as in ordinary 3D space. The last part is true, but the new inner product is actually different from the regular one in one very important respect: the value in equation (12.4) can be zero or even negative for nonzero four-vectors $\bar{\boldsymbol{x}}$! To see what's going on, we return to the spacetime diagram, in particular figure 12.1c. Suppose we have a particle traveling in the x-direction (taking $y = z = 0$ for convenience). What speed does it need for the length of its four-vector to vanish? For that to happen we need $ct = x$, or $v = x/t = c$, so $\bar{\boldsymbol{x}} \cdot \bar{\boldsymbol{x}}$ becomes zero for something traveling at the speed of light. Likewise, $\bar{\boldsymbol{x}} \cdot \bar{\boldsymbol{x}}$ is positive for a particle traveling slower than light, and negative for a particle traveling faster than light (which is of course impossible, since such a particle would need to first reach the speed of light, which as we've seen can never be done). However, we can consider the four-vector $\bar{\boldsymbol{x}}$ between any two points in spacetime, and from the sign of $\bar{\boldsymbol{x}} \cdot \bar{\boldsymbol{x}}$ tell whether they can be connected through regular (slower than light) travel, by a light beam, or not at all. The first we call *timelike*, the second *lightlike*, and the third *spacelike*:

$$\begin{array}{ll} \bar{\boldsymbol{x}} \cdot \bar{\boldsymbol{x}} > 0 & \text{timelike} \\ \bar{\boldsymbol{x}} \cdot \bar{\boldsymbol{x}} = 0 & \text{lightlike} \\ \bar{\boldsymbol{x}} \cdot \bar{\boldsymbol{x}} < 0 & \text{spacelike} \end{array} \tag{13.3}$$

Two events which are connected by a spacelike four-vector cannot influence each other: there is no way to send a signal between them, and therefore there is no way to transfer information.

13.2 LORENTZ TRANSFORMATION MATRIX AND METRIC TENSOR*

In this section we've joined space and time in a single four-vector, and defined a new inner product on the space of those four-vectors. In chapter 11 we defined the Lorentz transformations of the space and time coordinates, which are linear transformations. Linear transformations can of course be represented by matrices, and for our four-vectors, we can write down the appropriate Lorentz transformation matrix, rewriting equation (11.12) as a vector equation:

$$\bar{x}' = L\bar{x}. \tag{13.4}$$

Here L is a 4×4 matrix:

$$L = \begin{pmatrix} \gamma(u) & -\gamma(u)\frac{u}{c} & 0 & 0 \\ -\gamma(u)\frac{u}{c} & \gamma(u) & 0 & 0 \\ 0 & 0 & 1 & 0 \\ 0 & 0 & 0 & 1 \end{pmatrix}. \tag{13.5}$$

Like with the four-vectors, we start labeling the rows and columns of L with index 0. To indicate the difference with matrices in regular space, it is conventional to indicate indices of regular-space vectors and matrices with Roman letters (like v_i for the ith component of vector $\boldsymbol{v}$, and A_{ij} for the ith row, jth column of matrix A), and those of Minkowski-space vectors and matrices with Greek letters - so we write x_μ for the μth component of the four-vector $\bar{x}$, where μ can be 0, 1, 2, or 3.

We can also write equation (13.4) in index form:

$$x'_\mu = \sum_{\nu=0}^{3} L_\mu^\nu x_\nu. \tag{13.6}$$

The kind of summation in equation (13.6) is very common - so common, that most people adopt the *summation convention*: you always sum over repeated indices - from 1 to 3 if they're Roman, from 0 to 3 if they're Greek[1] - and leave out the explicit summation sign. Equation (13.6) then simply reads $x'_\mu = L_\mu^\nu x_\nu$. Some authors add the strict condition that you only sum over repeated indices if one is down and the other up (like in equation (13.6), where the up index of L_μ^ν and the down index of x_ν are summed over), whereas others sum over everything that's repeated, so you should always be explicit in which form of the convention you follow. Just to be clear: equation (13.6) is simply the expression in elements of the multiplication of the matrix given in equation (13.5) with the four-vector $\bar{x}$, so you could also write it as

$$\begin{pmatrix} x'_0 \\ x'_1 \\ x'_2 \\ x'_3 \end{pmatrix} = \begin{pmatrix} \gamma(u) & -\gamma(u)\frac{u}{c} & 0 & 0 \\ -\gamma(u)\frac{u}{c} & \gamma(u) & 0 & 0 \\ 0 & 0 & 1 & 0 \\ 0 & 0 & 0 & 1 \end{pmatrix} \begin{pmatrix} x_0 \\ x_1 \\ x_2 \\ x_3 \end{pmatrix}. \tag{13.7}$$

The Lorentz transform can be thought of as a *map* (in the mathematical sense[2]) from Minkowski space to itself: it takes a four-vector and maps it on another four-vector. The other operation we've encountered, the inner product of two four-vectors, is also a map, but its properties are different: it takes as its argument two four-vectors, and the result is a real number. Moreover, as with all inner products, the map is linear in both its arguments (we call such maps bilinear). A bilinear map from a vector space to a real number is a (rank) *two-tensor*[3]. The tensor that gives the inner product of two position vectors is known as the *metric* tensor, because this inner product is related to the vectors' length. Any two-tensor can be represented as a matrix; the metric tensor $g^{\mu\nu}$ of Minkowski space is given by:

$$g^{\mu\nu} = \begin{pmatrix} 1 & 0 & 0 & 0 \\ 0 & -1 & 0 & 0 \\ 0 & 0 & -1 & 0 \\ 0 & 0 & 0 & -1 \end{pmatrix}, \tag{13.8}$$

[1] In one of Einstein's most quoted statements, he jokingly called the summation convention 'his most important contribution to science'.

[2] If you've never heard of maps before: you're actually familiar with them, as any mathematical function is a map. For instance, the function $f(x,y,z) = \sqrt{x^2+y^2+z^2}$, which gives you the norm of the vector $\boldsymbol{x} = (x,y,z)$, is a map from $\mathbb{R}^3$ to $\mathbb{R}$.

[3] A vector is a (rank) one-tensor; a (rank) n-tensor maps n vectors on $\mathbb{R}$, linear in each of its arguments.

and we can use it to calculate the inner product as (using the summation convention on both indices):

$$\bar{\boldsymbol{a}} \cdot \bar{\boldsymbol{b}} = g^{\mu\nu} a_\mu b_\nu = a_0 b_0 - a_1 b_1 - a_2 b_2 - a_3 b_3. \tag{13.9}$$

The metric tensor may not seem very useful at this point - after all, we don't really need it to calculate the inner product, as long as we remember the definition (13.2). It will start to matter in the theory of general relativity, in which spacetime can be *curved* by gravity. In that case the components of the metric tensor change, and inner products of position four-vectors become complicated expressions.

13.3 VELOCITY AND MOMENTUM FOUR-VECTORS

We already encountered the position four-vector in section 13.1. The four components of this four-vectors represent a 'position' in the four-dimensional spacetime of special relativity: three coordinates in space, one in time. We defined a dot product for these four-vectors in equation (13.2), and an associated length of a four vector (the square root of the dot product with itself). Positions are however not the only vector quantities we know - we encountered many others in part I, like velocity and momentum. Like position, these vector quantities have four-vector versions in special relativity. Also, like the position four-vector, we'll demand that their length should be invariant under Lorentz transformations.

Let us start with velocity. In classical mechanics, we calculate velocity as the time derivative of the position: $\boldsymbol{v} = \mathrm{d}\boldsymbol{x}/\mathrm{d}t$. It is a straightforward exercise to check that the time derivative of the position four-vector $\bar{\boldsymbol{x}}$ is *not* a four-vector itself, as its length is not invariant under Lorentz transformations. Perhaps this does not surprise you: after all, t is the time as measured by an outside observer, and we've argued extensively that time depends on the observer. Instead of taking the derivative to t, we should therefore take the derivative with respect to time on a comoving clock, i.e., the proper time τ, in which case we do get the velocity four-vector:

$$\bar{\boldsymbol{v}} \equiv \frac{\mathrm{d}\bar{\boldsymbol{x}}}{\mathrm{d}\tau} = \gamma(v)(c, v_x, v_y, v_z), \tag{13.10}$$

where v_x, v_y and v_z are the classical velocity components, and v is the speed. A straightforward calculation shows that the length of $\bar{\boldsymbol{v}}$ equals c, which is certainly invariant under Lorentz transformations.

Momentum in classical mechanics is defined as mass times velocity. We use the same definition in special relativity, with the only difference that the velocity is now given by the velocity four-vector. Because the mass is just a scalar, it simply multiplies the velocity four-vector, and is not affected by Lorentz transformations, so this procedure again yields a proper four-vector with length mc:

$$\bar{\boldsymbol{p}} \equiv m\bar{\boldsymbol{v}} = m\gamma(v)(c, v_x, v_y, v_z) = (\gamma(v)mc, p_x, p_y, p_z), \tag{13.11}$$

where $p_x = \gamma(v)mv_x$, $p_y = \gamma(v)mv_y$ and $p_z = \gamma(v)mv_z$ are the components of the 'three-momentum' $\boldsymbol{p} = (p_x, p_y, p_z)$. Note that unlike the velocities, these components are not equal to their classical counterparts[4], as they differ by a factor $\gamma(v)$.

13.4 RELATIVISTIC ENERGY

The last three (or 'spatial') components of the momentum four-vector give us the regular components of the momentum, times the factor $\gamma(v)$. What about the zeroth (or 'temporal') component? To interpret it, we expand $\gamma(v)$, and find:

$$cp_0 = \gamma(v)mc^2 = \left[1 + \frac{1}{2}\left(\frac{v}{c}\right)^2 + \frac{3}{8}\left(\frac{v}{c}\right)^4 + \ldots\right] mc^2 = mc^2 + \frac{1}{2}mv^2 + \ldots \tag{13.12}$$

The second term in this expansion should be familiar: it's the kinetic energy of the particle. The third and higher terms are corrections to the classical kinetic energy - just like the higher-order terms in the spatial components are corrections to the classical momenta. The first term however is new: an extra energy contribution due to the mass of the particle. The whole term can now be interpreted as the *relativistic energy* of the particle:

$$E = \gamma(v)mc^2 = mc^2 + K. \tag{13.13}$$

[4] This difference is unfortunate, as it might be confusing. We keep the classical components of the velocity, as $\gamma(v)$ is expressed in these classical values. For the momenta however, we'll frequently use the relativistic version, so it is more convenient to define p_x, p_y and p_z as components of $\bar{\boldsymbol{p}}$.

We can now write the zeroth component of the momentum four-vector as $p_0 = E/c$. Based on this interpretation, the four-vector is sometimes referred to as the energy-momentum four-vector.

A very useful relation can now easily be derived by calculating the length of the energy-momentum four-vector in two ways. On the one hand, it's given by (leaving out the square root for convenience)

$$\bar{\boldsymbol{p}} \cdot \bar{\boldsymbol{p}} = m^2 \bar{\boldsymbol{v}} \cdot \bar{\boldsymbol{v}} = m^2 c^2, \tag{13.14}$$

while on the other hand, we could also simply expand in the components of $\bar{\boldsymbol{p}}$ itself to get:

$$\bar{\boldsymbol{p}} \cdot \bar{\boldsymbol{p}} = \left(\frac{E}{c}\right)^2 - \boldsymbol{p} \cdot \boldsymbol{p}, \tag{13.15}$$

where $\boldsymbol{p}$ is again the spatial part of $\bar{\boldsymbol{p}}$. Combining equations (13.14) and (13.15), we get:

$$\boxed{E^2 = m^2 c^4 + p^2 c^2,} \tag{13.16}$$

where $p^2 = \boldsymbol{p} \cdot \boldsymbol{p}$. Equation (13.16) is the general form of Einstein's famous formula $E = mc^2$, to which it reduces for stationary particles (i.e., when $v = p = 0$).

13.5 CONSERVATION OF ENERGY AND MOMENTUM

In classical mechanics, energy and momentum were separate entities, each obeying its own conservation law. In special relativity, they are two parts of the same quantity (the energy-momentum four-vector), just like time and space are two parts of the same position four-vector. Consequently, energy and momentum have to obey the same rules in special relativity. Fortunately, a conservation law on a vector quantity applies to each of its components, and so conservation of energy and momentum translates to conservation of the energy-momentum four-vector $\bar{\boldsymbol{p}}$. However, unlike in classical mechanics, mass is no longer conserved: since it is now interpreted as a part of the total energy of a system (equation 13.13), it can be converted into or created from kinetic energy. The equivalence of mass and energy has important consequences for collision experiments, including a whole new type of 'collisions': radioactive decay of matter.

You might complain that we haven't actually proved that the energy-momentum four-vector is conserved in special relativity (and you would be right). What we have done is define the relativistic energy $E = \gamma(v)mc^2$ and three-momentum $\boldsymbol{p} = \gamma(v)m\boldsymbol{v}$, as well as the energy-momentum four-vector $\bar{\boldsymbol{p}}$. We have also shown that with these definitions, $\bar{\boldsymbol{p}}$ is a proper four-vector, in the sense that it is invariant under Lorentz transformations. Therefore, we know that if it is conserved in one inertial frame, it must also be conserved in all others. We also know that our relativistic energy and momentum revert to the classical kinetic energy (plus a constant, mc^2) and the classical momentum mv at low velocities. The conservation laws for these classical quantities follow from Newton's second and third laws of motion, respectively. In special relativity, we no longer take these laws as our axioms, only retaining Newton's first law of motion in inertial reference frame. We therefore *cannot* prove conservation of the energy-momentum four-vector mathematically, and must take it as an axiom. As I've just argued, this axiom is consistent with the laws of classical mechanics in the low-velocity limit. It is also consistent with experimental data - which, like Einstein's postulates and Newton's laws in classical mechanics, is the ultimate test of our physical model.

13.6 PROBLEMS

13.1 In high-energy physics, it is customary to express the mass of elementary particles not in kilograms but in MeV/c^2, expressing the fact that (rest) mass is a form of energy. An MeV or mega-electron-Volt is one million (the 'mega') times the (kinetic) energy an electron gains when it moves through an electric field between two positions with an electric potential difference of 1 volt, or 1 joule per coulomb. One electron-volt thus corresponds to an amount of energy (in joules) equal in number to the charge of the electron in coulombs. Express the mass of both the electron and the proton in MeV/c^2; you may find the numbers in table B.1 useful.

13.2 For an arbitrary particle of (rest) mass m, find the speed at which its kinetic energy equals its rest energy.

13.3 We constructed four vectors in such a way that their length is invariant under Lorentz transformations. The length of a four-vector is defined as the square root of its dot product with itself: $|\bar{\boldsymbol{x}}| = \sqrt{\bar{\boldsymbol{x}} \cdot \bar{\boldsymbol{x}}} = x_0^2 - x_1^2 - x_2^2 - x_3^2$. In equation (13.2) we also defined the dot product of two arbitrary four-vectors $\bar{\boldsymbol{a}}$ and $\bar{\boldsymbol{b}}$.

(a) Show that the sum of two four-vectors is again a four-vector (i.e., show that the length is invariant under Lorentz transformations, and the components transform the same way that those of the position four-vector do).

(b) Calculate the square of the length of the four-vectors $\bar{\boldsymbol{a}} + \bar{\boldsymbol{b}}$ and $\bar{\boldsymbol{a}} - \bar{\boldsymbol{b}}$.

(c) Use your answer at (b) to write the dot product of $\bar{\boldsymbol{a}}$ and $\bar{\boldsymbol{b}}$ as a linear combination of quantities that are invariant under Lorentz transformations (thus showing that the dot product is also invariant).

13.4 A particle with mass m has three-momentum $\boldsymbol{p}$ as measured in an inertial lab frame S. Find the particle's energy as measured by an observer with three-velocity $\boldsymbol{u}$. *Hint*: Determine the four-vectors of the particle's momentum and the observer's motion both in the lab frame S and the observer's rest frame S', then use the fact that inner products of four-vectors are invariant under Lorentz transformations.

14

RELATIVISTIC COLLISIONS

In supercolliders such as the ones at CERN in Geneva and (formerly) Fermilab in Chicago, small particles like electrons and protons are accelerated to speeds near that of light, then made to collide with each other in an attempt to create exotic types of matter (i.e., non-common particles). The very reason why this can be done is the relation between energy, mass and momentum given by the general version of Einstein's famous equation, equation (13.16). This equation tells us that if the incoming particles have sufficiently high kinetic energy, we can create new particles with more mass than the originals had. The process by which this happens is the realm of quantum field theory, but the mechanics of the collisions can be studied within special relativity.

Just like in classical mechanics, we can define a *totally inelastic collision* as any collision in which the particles stick together. We define a *totally elastic collision* as a collision in which the momentum, kinetic energy, *and* mass of all particles are conserved. We'll have one more type, that has no classical counterpart: *radioactive decay*, in which a particle falls apart into multiple particles - a sort of time-reversed inelastic collision. All cases can be analyzed using conservation of energy-momentum. Although that basic concept is in principle sufficient, there are many cases for which writing out the components of the energy-momentum four vector as four equations is not the easiest way to find (say) the energies or momenta of the outgoing particles. There are some other tricks that you can use - in particular the invariance of the length of the energy-momentum four vector, both in a collision process and under a Lorentz transformation. A few examples will help to illustrate this point.

14.1 PHOTONS

Before we dive into the examples, there's one particle that requires special attention: the photon, or quantum of light - from that other early-20th-century theory known as quantum mechanics. Photons travel (by definition) at the speed of light, and need therefore be massless. They do carry energy though, which is related to their frequency f (or wavelength λ, or color) through

$$E_{\text{photon}} = hf = \frac{hc}{\lambda}, \tag{14.1}$$

where h is Planck's constant. Since photons have nonzero energy, they also have nonzero momentum through Einstein's equation (13.16), despite the fact that they have no mass[1]

$$p_{\text{photon}} = \frac{E_{\text{photon}}}{c} = \frac{hf}{c} = \frac{h}{\lambda}. \tag{14.2}$$

A photon with frequency f (and thus energy $E = hf$) traveling in the positive x direction has a very simple energy-momentum four vector:

$$\bar{\boldsymbol{p}}_{\text{photon}} = (E/c, E/c, 0, 0). \tag{14.3}$$

The length of this four-vector, unsurprisingly, is zero.

[1]Note that relativistic momentum is given by $p = \gamma(v)mv$; substituting $v = c$ gives $\gamma(c) = \infty$, and so this expression gives us the product of infinity with zero for the momentum of the photon - that could be anything, and thus is meaningless. The way to calculate the momentum of the photon is through equation (14.2). Although the photon momentum is small, it is large enough to be measured, and even useful in devices known as optical tweezers, where focused laser beams are used to move micron-sized objects around.

14.2 TOTALLY INELASTIC COLLISION

In a totally inelastic collision, particles stick together. A possible example is the absorption of a photon by a massive particle, resulting in an increase in its mass, as well as possibly a change in its momentum. Let's consider, as an example, a particle of mass m that is initially at rest, and absorbs an incoming photon with energy E_γ. There are now three ways to calculate the energy and momentum of the particle after this collision.

Method 1: We have conservation of both (total) energy and momentum. Before the collision, the massive particle has energy $E_\mathrm{i} = mc^2$ (as it is standing still), and the total energy of the system is $E_\gamma + mc^2$, which must be conserved. The total energy of the particle after the collision is $E_\mathrm{f} = \gamma(v) m_\mathrm{f} c^2$, where both the velocity v and the mass m_f are unknown. The total momentum before the collision is E_γ / c, as the particle is initially standing still (and thus has momentum zero), while after the collision it is $\gamma(v) m_\mathrm{f} v$. We thus have:

$$E_\gamma + mc^2 = \gamma(v) m_\mathrm{f} c^2, \tag{14.4}$$

$$E_\gamma = \gamma(v) m_\mathrm{f} v c. \tag{14.5}$$

We thus have two equations with two unknowns (v and m_f). If we divide equation (14.5) by (14.4), we get an expression for the final velocity v, which we can substitute back in either equation to solve for m_f (and potentially use to calculate the momentum after the collision). This is not pretty though, as we'll have complicated factors due to the presence of $\gamma(v)$.

Method 2: The four-momentum of the system is conserved during the collision. We have

$$\bar{\boldsymbol{p}}_\gamma = \left(\frac{E_\gamma}{c}, \frac{E_\gamma}{c}, 0, 0\right) \quad \text{for the photon,} \tag{14.6}$$

$$\bar{\boldsymbol{p}}_\mathrm{i} = (mc, 0, 0, 0) \quad \text{for the massive particle before the collision,} \tag{14.7}$$

$$\bar{\boldsymbol{p}}_\mathrm{f} = \left(\frac{E_\mathrm{f}}{c}, p_\mathrm{f}, 0, 0\right) \quad \text{for that particle after the collision.} \tag{14.8}$$

From $\bar{\boldsymbol{p}}_\gamma + \bar{\boldsymbol{p}}_\mathrm{i} = \bar{\boldsymbol{p}}_\mathrm{f}$ we can read off two equations:

$$E_\gamma + mc^2 = E_\mathrm{f}, \tag{14.9}$$

$$E_\gamma / c = p_\mathrm{f}, \tag{14.10}$$

which immediately give us the final energy and momentum in terms of the initial ones. We can now find the final mass through Einstein's equation (13.16):

$$m_\mathrm{f}^2 c^4 = E_\mathrm{f}^2 - p_\mathrm{f}^2 c^2 = (E_\gamma + mc^2)^2 - E_\gamma^2 = (E_\gamma + mc^2) mc^2. \tag{14.11}$$

This approach circumvents the use of the $\gamma(v)$ factor, because we only use energy and momentum, not (classical) velocity. If we now want the velocity, we could still calculate it from the combination of m_f and either E_f or p_f, but since it were the mass and momentum we were after, there's no need to do so.

Method 3: Since the total energy-momentum four vector is conserved in the collision, so must be its length (or the square of the length), which is trivial to calculate (remember that $\bar{\boldsymbol{p}} \cdot \bar{\boldsymbol{p}} = m^2 c^2$). We can often exploit this fact to make the maths much simpler. To see how this works, let's consider the full four-vector equation for our example: $\bar{\boldsymbol{p}}_\gamma + \bar{\boldsymbol{p}}_\mathrm{i} = \bar{\boldsymbol{p}}_\mathrm{f}$, so

$$(\bar{\boldsymbol{p}}_\gamma + \bar{\boldsymbol{p}}_\mathrm{i}) \cdot (\bar{\boldsymbol{p}}_\gamma + \bar{\boldsymbol{p}}_\mathrm{i}) = \bar{\boldsymbol{p}}_\mathrm{f} \cdot \bar{\boldsymbol{p}}_\mathrm{f} \tag{14.12}$$

$$\bar{\boldsymbol{p}}_\gamma \cdot \bar{\boldsymbol{p}}_\gamma + \bar{\boldsymbol{p}}_\mathrm{i} \cdot \bar{\boldsymbol{p}}_\mathrm{i} + 2\bar{\boldsymbol{p}}_\gamma \cdot \bar{\boldsymbol{p}}_\mathrm{i} = \bar{\boldsymbol{p}}_\mathrm{f} \cdot \bar{\boldsymbol{p}}_\mathrm{f} \tag{14.13}$$

$$0 + m^2 c^2 + 2E_\gamma m = m_\mathrm{f}^2 c^2, \tag{14.14}$$

which immediately gives us m_f. If we also want E_f or p_f, we can again use equations (14.9) and (14.10) for the components, but if we only wanted the final mass, we're done in one step.

Note that although method 3 usually is the easiest route to your answer, it isn't always - and it is a good idea to at least be aware of the other options.

14.3 Radioactive decay and the center-of-momentum frame

Radioactive decay is the process by which unstable particles with high mass fall apart into more stable particles with lower mass. Although the process itself is quantum mechanical in nature, the dynamics of radioactive decay are described by special relativity, and are essentially identical to those of an inelastic collision in reverse.

Decay may occur spontaneously (as a random process), but can also be stimulated, by the absorption of a (typically small, e.g. a photon or electron) particle by the unstable one - a process used in nuclear reactors. Because the absorbed particle also carries energy, in stimulated decay the masses of the resulting particles can add up to something more than the rest mass of the original particle. An important question in nuclear physics is what the *threshold energy* of a given reaction is, i.e., the minimum energy the incoming particle must have for the process to be possible. This is not simply the differences in the mass energy of the original and the resulting particles, as in the collision process momentum must also be conserved. To illustrate how to approach such a problem, let's again consider a concrete example: the threshold energy for the reaction in which a proton ($m_p = 938\,\mathrm{MeV}/c^2$), initially at rest, absorbs a photon, and then emits a neutral pion ($m_\pi = 135\,\mathrm{MeV}/c^2$), see figure 14.1a.

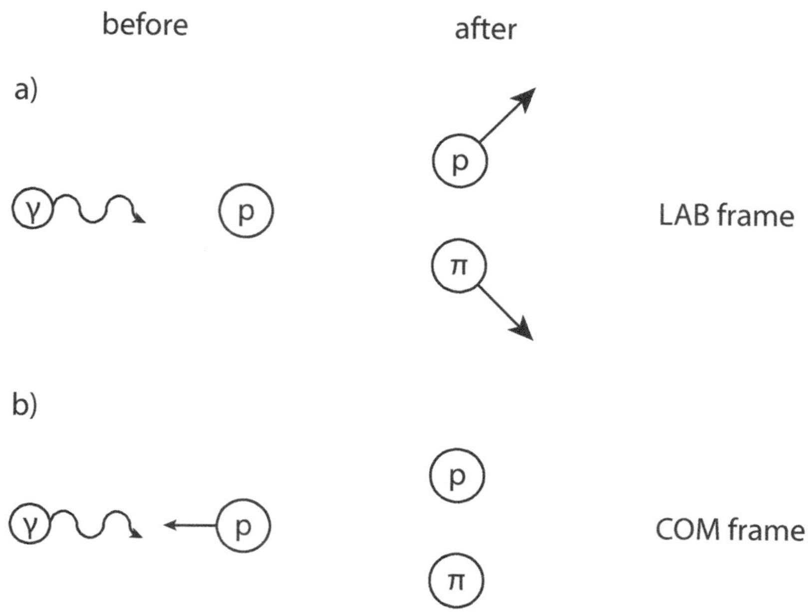

Figure 14.1: Example of stimulated radioactive decay: a proton, initially at rest, absorbs a photon, and then emits a neutral pion. The reaction is shown in the lab frame in (a), and in the center-of-momentum frame in (b).

Figuring out what the minimal required energy is in the lab frame is not easy, as you have to account for the kinetic energy of the particles after the reaction. There is however a system in which the reaction products are standing still: the *center-of-momentum* frame, the relativistic analog of the center-of-mass frame of classical mechanics[2]. The center-of-momentum frame is defined as the frame in which the total momentum of all particles adds up to zero. In our specific example, before the collision, only the photon carries a momentum, equal to its energy E_γ divided by the speed of light. In general, the total momentum in the system can be a three-vector, equal to $\boldsymbol{p}_\mathrm{T} = \sum_i \boldsymbol{p}_i$, while the total energy is given by $E_\mathrm{T} = \sum_i E_i$. If we choose our coordinates such that the x-direction coincides with that of $\boldsymbol{p}_\mathrm{T}$, the energy-momentum four-vector of the entire system becomes $\bar{\boldsymbol{p}}_\mathrm{T} = (E_\mathrm{T}/c, p_\mathrm{T}, 0, 0)$, where $p_\mathrm{T} = |\boldsymbol{p}_\mathrm{T}|$. If we go to any different inertial frame S' moving with velocity v in the positive x direction, the components of the energy-momentum four vector are given by the Lorentz transform of $\bar{\boldsymbol{p}}_\mathrm{T}$:

$$\bar{\boldsymbol{p}}'_\mathrm{T} = \gamma(v)\left(\frac{E_\mathrm{T}}{c} - \frac{v}{c}p_\mathrm{T}, p_\mathrm{T} - \frac{v}{c}\frac{E_\mathrm{T}}{c}, 0, 0\right), \tag{14.15}$$

[2] As our system includes a photon, a center-of-mass frame doesn't make sense here, as the photon has no mass - but it has nonzero momentum, so we can make a transformation to a system in which the total momentum vanishes.

so we end up in a frame in which the total momentum is zero if we pick

$$v_{\mathrm{COM}} = \frac{c^2 p_{\mathrm{T}}}{E_{\mathrm{T}}} \tag{14.16}$$

for our velocity. In particular, we see that we can always make this transformation, and that the center-of-momentum frame is an inertial frame.

Back to our example: why do we care? The answer is almost tautological: if the total momentum is zero before the collision, it is also zero afterwards - and so in the COM frame, the particles can all be standing still (see figure 14.1b). That certainly corresponds to the lowest possible kinetic energy of the system, so the energy of the incoming photon is all converted to mass - and that must thus be the threshold energy we're looking for. Interestingly, to answer our original question, we don't even need to calculate what the actual velocity of the COM frame is, just the fact that it exists is sufficient. In the COM frame, we have, by conservation of four-momentum:

$$\bar{\boldsymbol{p}}'_{\gamma} + \bar{\boldsymbol{p}}'_{\mathrm{p,i}} = \bar{\boldsymbol{p}}'_{\mathrm{p,f}} + \bar{\boldsymbol{p}}'_{\pi}, \tag{14.17}$$

and therefore also

$$(\bar{\boldsymbol{p}}'_{\gamma} + \bar{\boldsymbol{p}}'_{\mathrm{p,i}})^2 = (\bar{\boldsymbol{p}}'_{\mathrm{p,f}} + \bar{\boldsymbol{p}}'_{\pi})^2 = (m_{\mathrm{p}}c + m_{\pi}c)^2, \tag{14.18}$$

where the last equality follows from the fact that the reactants are standing still. Now the left-hand-side of equation (14.18) is the length of a four-vector, and we've proven that these lengths are invariant under Lorentz transformations - so its value is equal to that of $(\bar{\boldsymbol{p}}_{\gamma} + \bar{\boldsymbol{p}}_{\mathrm{p,i}})^2$ in the lab frame. In that frame, we have $\bar{\boldsymbol{p}}_{\gamma} = (E_{\gamma}/c)(1,1,0,0)$ and $\bar{\boldsymbol{p}}_{\mathrm{p,i}} = (m_{\mathrm{p}}c,0,0,0)$, so we end up with an easy equation form E_{γ}:

$$(m_{\mathrm{p}} + m_{\pi})^2 c^2 = \bar{\boldsymbol{p}}_{\gamma}^2 + \bar{\boldsymbol{p}}_{\mathrm{p,i}}^2 + 2\bar{\boldsymbol{p}}_{\gamma} \cdot \bar{\boldsymbol{p}}_{\mathrm{p,i}} = 0 + m_{\mathrm{p}}^2 c^2 + 2E_{\gamma} m_{\mathrm{p}}, \tag{14.19}$$

or

$$E_{\gamma} = \frac{m_{\pi}^2 + 2m_{\pi} m_{\mathrm{p}}}{2m_{\mathrm{p}}} c^2 = 145\,\mathrm{MeV}. \tag{14.20}$$

In this example, we thus need at least 10 MeV of energy more than the mass of the particle we've created.

Note that in finding the threshold energy in the example, we again heavily relied on the four-vector properties of $\bar{\boldsymbol{p}}$ - not only its length (like in the third method of section 14.2), but also the invariance of that length under Lorentz transformations. Using these properties results in easy equations to solve, while if you'd ignore them, you'd probably get stuck trying to figure out what the kinetic energy of the reaction products is.

14.4 TOTALLY ELASTIC COLLISION: COMPTON SCATTERING

As a final example of a collision in special relativity, we consider the totally elastic case: a collision in which the total momentum, total kinetic energy, and the mass of all particles are conserved. An example of such a collision is *Compton scattering*: the collision between a photon and an electron, resulting in a transfer of energy from one to the other, visible in a change of wavelength of the photon. For our example, we'll take the electron to be initially stationary, and the photon to be coming in along the x-axis; after the collision both particles have nonzero velocities in both the x and y directions (see figure 14.2).

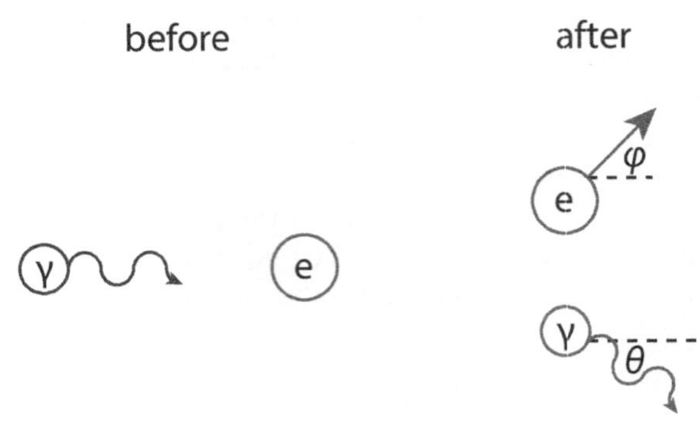

Figure 14.2: Compton scattering between a photon and an electron, resulting in a transfer of energy of the photon to the electron, measurable as a change in the photon's wavelength.

The four-momenta of the electron and photon before and after the collision are given by:

$$\bar{\boldsymbol{p}}_{\mathrm{e,i}} = \begin{pmatrix} m_{\mathrm{e}}c \\ 0 \\ 0 \\ 0 \end{pmatrix}, \quad \bar{\boldsymbol{p}}_{\gamma,\mathrm{i}} = \frac{E_{\mathrm{i}}}{c}\begin{pmatrix} 1 \\ 1 \\ 0 \\ 0 \end{pmatrix}, \quad \bar{\boldsymbol{p}}_{\mathrm{e,f}} = \begin{pmatrix} E_{\mathrm{e,f}}/c \\ p_{\mathrm{e,f}}\cos\phi \\ p_{\mathrm{e,f}}\sin\phi \\ 0 \end{pmatrix}, \quad \bar{\boldsymbol{p}}_{\gamma,\mathrm{f}} = \frac{E_{\mathrm{f}}}{c}\begin{pmatrix} 1 \\ \cos\theta \\ -\sin\theta \\ 0 \end{pmatrix}. \tag{14.21}$$

We can now solve for the energy E_{f} of the outgoing photon (and thus its wavelength) in terms of that of the incoming photon (E_{i}) and the scattering angle θ. There are again (at least) two ways to do this. One is to compare the components of the initial and final energy-momentum four-vector term by term. The other is to again use the fact that we know about the length of the four-vector to immediately eliminate the scattering angle ϕ of the electron. To do so, we first rewrite the conservation of energy-momentum equation, $\bar{\boldsymbol{p}}_{\mathrm{e,i}}+\bar{\boldsymbol{p}}_{\gamma,\mathrm{i}} = \bar{\boldsymbol{p}}_{\mathrm{e,f}} + \bar{\boldsymbol{p}}_{\gamma,\mathrm{f}}$ to isolate the term of the outgoing electron, and then take the square, to get:

$$(\bar{\boldsymbol{p}}_{\mathrm{e,i}} + \bar{\boldsymbol{p}}_{\gamma,\mathrm{i}} - \bar{\boldsymbol{p}}_{\gamma,\mathrm{f}})^2 = \bar{\boldsymbol{p}}_{\mathrm{e,f}}^2 \tag{14.22}$$

$$\bar{\boldsymbol{p}}_{\mathrm{e,i}}^2 + \bar{\boldsymbol{p}}_{\gamma,\mathrm{i}}^2 + \bar{\boldsymbol{p}}_{\gamma,\mathrm{f}}^2 + \bar{\boldsymbol{p}}_{\mathrm{e,i}} \cdot \bar{\boldsymbol{p}}_{\gamma,\mathrm{i}}^2 - 2\bar{\boldsymbol{p}}_{\mathrm{e,i}} \cdot \bar{\boldsymbol{p}}_{\gamma,\mathrm{f}} - 2\bar{\boldsymbol{p}}_{\gamma,\mathrm{i}} \cdot \bar{\boldsymbol{p}}_{\gamma,\mathrm{f}} = \bar{\boldsymbol{p}}_{\mathrm{e,f}}^2 \tag{14.23}$$

$$m_e^2c^2 + 0 + 0 + 2m_{\mathrm{e}}E_{\mathrm{i}} - 2m_{\mathrm{e}}E_{\mathrm{f}} - 2\frac{E_{\mathrm{i}}E_{\mathrm{f}}}{c^2}(1-\cos\theta) = m_e^2c^2, \tag{14.24}$$

from which we can solve for E_{f}. Rewriting to wavelengths (through $E = hf = hc/\lambda$), we get

$$\lambda_{\mathrm{f}} = \lambda_{\mathrm{i}} + \frac{h}{m_{\mathrm{e}}c}(1-\cos\theta). \tag{14.25}$$

14.5 PROBLEMS

14.1 A photon with frequency f collides with a stationary atom with rest mass m. In the collision, the photon is absorbed by the atom. Determine the mass and speed of the atom after the collision.

14.2 A particle with mass m and kinetic energy $2mc^2$ collides with a stationary particle with mass $2m$. After the collision, the two particles are fused into a single particle. Find both the mass and the speed of this new particle.

14.3 A stationary atomic nucleus undergoes a radioactive process known as β-decay, in which one of its neutrons (with rest mass $m_n = 939.6$ MeV) falls apart into a proton (which remains in the nucleus, rest mass $m_p = 938.3$ MeV), an electron (rest mass $m_e = 0.5$ MeV), and an anti-neutrino. Neutrino's are very light particles; we'll take the emitted neutrino to be effectively massless and thus travel at the speed of light with momentum p_ν. The nucleus remains stationary. Find the momenta p_ν and p_e of the emitted neutrino and electron, as well as the speed of the emitted electron.

14.4 A proton with rest mass m_p and momentum p_p is moving in the positive x-direction. A photon with frequency f is traveling in the negative x-direction, and collides head-on with the proton. After the collision, both proton and photon are traveling in the positive x-direction. Show that the frequency f' of the photon after the collision is given by

$$f' = \frac{E_p + cp_p}{E_p - cp_p + 2hf} f,$$

where E_p is the energy of the proton before the collision.

14.5 Particles like the electrons in atomic orbitals can be in a low-energy ground state (with energy E_0), or, by absorbing a photon, be put in a higher-energy excited state (with energy E_1). The particle can return to the ground state by emitting another photon. Quantum mechanics tells us that only very specific states with very specific, discrete (or 'quantized') energies, are allowed.

(a) If the particle is initially at rest, the energy of an incoming photon with frequency ν (and energy $E = h\nu$) has to be slightly larger than the energy difference $\Delta E = E_1 - E_0$ between the particle's ground and exited states if the particle is to absorb the photon. Explain why.

(b) Show that for an incoming photon that is absorbed by an initially stationary particle, we have

$$h\nu_a = \Delta E\left(1 + \frac{\Delta E}{2E_0}\right).$$

(c) Likewise, show that for an initially stationary particle in the excited state with energy E_1, the energy of the emitted photon is given by

$$h\nu_e = \Delta E\left(1 - \frac{\Delta E}{2E_0}\right).$$

(d) Suppose we have two identical atoms, one of which contains an electron in the exited state, and the other only electrons in the ground state. The atom with the electron in the exited state emits a photon. Is there a possible scenario in which the other atom absorbs the photon (resonant absorption)?

14.6 **Matter-antimatter annihilation and creation** As you may have heard, for every elementary particle of 'ordinary' matter, there exists an antiparticle of 'antimatter', which shares many characteristics with its ordinary counterpart (such as the mass), whereas others are opposite (such as the charge). When a particle and its antiparticle meet, they completely annihilate, converting all of their combined mass into pure energy, in the form of radiation (i.e. photons). The most common antiparticle is that of the electron, which is known as the positron. First, we consider an experiment in which an ordinary electron of mass m_e with momentum p_e hits a positron (mass m_e) at rest, at which point the two annihilate, producing two photons.

(a) Argue why such an annihilation must produce at least two photons.

(b) One of the two produced photons emerges at an angle of 60° to the direction of the incident electron. What is its energy?

(c) Find the angle (with the direction of the incident electron) at which the other photon emerges.

The opposite of annihilation, spontaneous creation of matter, can also happen: then a photon spontaneously converts to a particle-antiparticle pair.

(d) Why must the photon convert to a particle-antiparticle pair, rather than simply convert to a single particle?

(e) Find the minimum wavelength a photon must have to create an electron-positron pair. Where is this photon in the electromagnetic spectrum?

14.7 A pion (rest mass 140 MeV/c^2) with a momentum of 25 MeV/c decays into a muon (rest mass 106 MeV/c^2) and a neutrino (rest mass approximately zero; taken zero in this problem).

(a) Show that

$$\bar{\boldsymbol{p}}_\pi \cdot \bar{\boldsymbol{p}}_\mu = \frac{1}{2}(m_\pi^2 + m_\mu^2)c^2,$$

where $\bar{\boldsymbol{p}}_\pi$ and $\bar{\boldsymbol{p}}_\mu$ are the energy-momentum four-vectors of the pion and the muon.

(b) Determine the kinetic energy of the outgoing muon for the case that it travels in a direction that makes a 90° angle with that of the incoming pion.

(c) Again for the case that the direction of the outgoing muon makes a 90° angle with that of the incoming pion, find the direction of the outgoing neutrino.

(d) Finally, again for the same case, find the magnitude of the (three-)momentum of the outgoing neutrino. Express your answer in the same units as the given values (MeV/c)

14.8 A neutral kaon (or K meson) with a mass of 498 MeV/c^2 and an initial velocity of $c/2$ decays into two pions (one with a positive and one with a negative charge), each of which has a mass of 135 MeV/c^2.

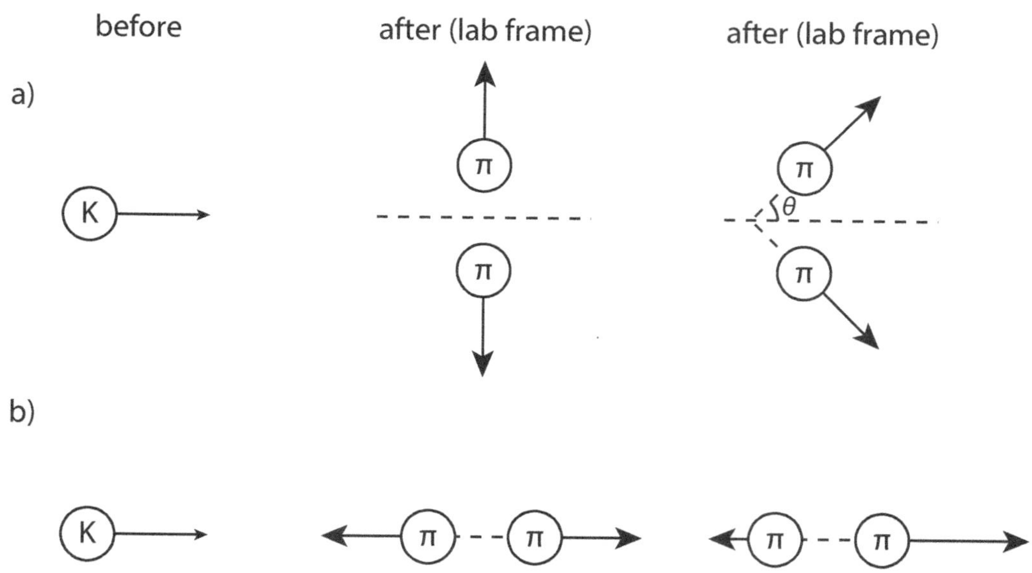

(a) Find the speeds and the angles of the pions in the lab frame if, in the rest frame of the kaon, they are emitted in opposite directions, whose line makes an angle of 90° with the propagation direction of the kaon?

(b) Answer the same question as in (a), for the case that the pions are emitted one in the same and one in the opposite direction as the kaon.

(c) Sometimes a kaon decays into more than two pions (there are also neutral pions; the charges of course need to add up to the kaon charge). Determine the maximum number of pions that our kaon can decay into.

(d) Prove that in any situation, the trajectories of the created pions are in one plane. *Hint*: do this in the kaon's rest frame first.

15

RELATIVISTIC FORCES AND WAVES

15.1 THE FORCE FOUR-VECTOR

As we've discussed in chapter 14 above, you can analyze all types of collisions in special relativity without ever making a reference to the forces they exert on each other. In fact, we haven't talked about force at all so far, and there's a good reason for that: forces, already frequently less practical than energies in classical mechanics, become veritable nightmares in special relativity. Nonetheless, there are some questions you can only answer with reference to forces - for example, what velocity a particle will get if you exert a certain force on it for a given period of time.

In classical mechanics, Newton's second law relates momenta and forces, through the time derivative of the momentum. In relativity, we'll therefore simply define the force four-vector as the derivative of the energy-momentum four vector with respect to the proper time (which gives a four-vector, as you can check easily):

$$\bar{\boldsymbol{F}} = \frac{\mathrm{d}\bar{\boldsymbol{p}}}{\mathrm{d}\tau} = \gamma(v)\left(\frac{1}{c}\frac{\mathrm{d}E}{\mathrm{d}t}, F_x, F_y, F_z\right). \tag{15.1}$$

We define the components of the three-force[1] $\boldsymbol{F}$ as the ('regular' or 'coordinate') time derivative of the three-momentum: $\boldsymbol{F} = \mathrm{d}\boldsymbol{p}/\mathrm{d}t$, so Newton's second law holds as long as you don't change your frame of reference. Likewise, Newton's third law holds, if you consider the three-force $\boldsymbol{F}$ in a fixed frame of reference. The zeroth term of $\bar{\boldsymbol{F}}$ contains the time derivative of the energy, which we defined as the power in section 3.1: $P = \mathrm{d}E/\mathrm{d}t$, again within the context of a fixed frame of reference.

There is a classical result that involves the force that does translate to special relativity for arbitrary reference frames: the work-energy theorem. To see how that comes about, consider a Lorentz transform from a comoving system (or instantaneous inertial frameS') to an arbitrary inertial frame S. In S', $\gamma(u) = \gamma(0) = 1$, so the space components of the force four-vector are just the components of the force three-vector (and Newton's second law holds); moreover, in this frame, $\mathrm{d}E'/\mathrm{d}t = \mathrm{d}(m/\sqrt{1-(u')^2/c^2})/\mathrm{d}t = 0$, because the derivative contains a factor u', which (by choice of frame) is zero. We thus have $\bar{\boldsymbol{F}}' = (0, F_x', F_y', F_z')$. The force is a four-vector, and therefore transforms according to equation (13.7):

$$\bar{\boldsymbol{F}} = \begin{pmatrix} F_0 \\ F_1 \\ F_2 \\ F_3 \end{pmatrix} = \begin{pmatrix} \gamma(u) & \gamma(u)\frac{u}{c} & 0 & 0 \\ \gamma(u)\frac{u}{c} & \gamma(u) & 0 & 0 \\ 0 & 0 & 1 & 0 \\ 0 & 0 & 0 & 1 \end{pmatrix} \begin{pmatrix} 0 \\ F_x' \\ F_y' \\ F_z' \end{pmatrix} = \begin{pmatrix} \gamma(u)\frac{u}{c}F_x' \\ \gamma(u)F_x' \\ F_y' \\ F_z' \end{pmatrix}, \tag{15.2}$$

so, comparing the components of $\bar{\boldsymbol{F}}$ in equations (15.1) and (15.2), we get

$$\frac{\mathrm{d}E}{\mathrm{d}t} = uF_x', \quad F_x = F_x', \quad F_y = \frac{F_y'}{\gamma(u)}, \quad F_z = \frac{F_z'}{\gamma(u)}. \tag{15.3}$$

The longitudinal force is thus the same in both frames, but the transversal force is not! Forces thus behave differently than you might naively expect under Lorentz transformations. Moreover, the transformation is *not*

[1] Some authors use $\boldsymbol{f}$ to avoid confusion with the four-force $\bar{\boldsymbol{F}}$; others use $\boldsymbol{F}$ for the three-force and $\bar{\boldsymbol{K}}$ for the four-force.

symmetric: we don't get $F'_y = F_y/\gamma(-u)$ (which would indeed directly contradict equation (15.2). The reason why we've lost this symmetry is that for forces, there is a special frame: that of the particle (here S'), where Newton's second law holds. In all other frames, we have to transform the forces according to equation (15.3).

There is a silver lining: the zero component in equation (15.2) gives us that $\mathrm{d}E = uF'_x\,\mathrm{d}t = F'_x\,\mathrm{d}x = F_x\,\mathrm{d}x$, which integrated gives the work-energy theorem: $\Delta E = F\Delta x = \Delta W$. As long as we stay away from the forces, work and energy will behave as we've come to expect.

15.2 The four-acceleration

We can of course also define a four-vector version of the acceleration, by taking the derivative of the four-velocity with respect to the proper time. As with the forces, we'll see that we're in for some nasty surprises, this time because the proper time derivative acts on the $\gamma(u)$ factor in the velocity as well as on the components:

$$\bar{\boldsymbol{a}} \equiv \frac{\mathrm{d}\bar{\boldsymbol{v}}}{\mathrm{d}\tau} = \gamma(v)\frac{\mathrm{d}}{\mathrm{d}t}\gamma(v)(c, v_x, v_y, v_z) = \left(\gamma^4(v)\frac{\boldsymbol{a}\cdot\boldsymbol{v}}{c}, \gamma^2(v)\boldsymbol{a} + \gamma^4(v)\frac{\boldsymbol{a}\cdot\boldsymbol{v}}{c^2}\boldsymbol{v}\right), \tag{15.4}$$

where we used the time derivative of the $\gamma(u)$ function

$$\frac{\mathrm{d}\gamma}{\mathrm{d}t} = \frac{\mathrm{d}}{\mathrm{d}t}\frac{1}{\sqrt{1-(v/c)^2}} = -\frac{1}{2}\frac{1}{(1-(v/c)^2)^{3/2}}\cdot\left(-2\frac{\boldsymbol{v}}{c^2}\cdot\frac{\mathrm{d}\boldsymbol{v}}{\mathrm{d}t}\right) = \gamma^3(v)\frac{\boldsymbol{v}\cdot\boldsymbol{a}}{c^2}, \tag{15.5}$$

and we've introduced the (classical) acceleration three-vector as the coordinate time derivative of the velocity three-vector: $\boldsymbol{a} = \mathrm{d}\boldsymbol{v}/\mathrm{d}t$. As you can see in equation (15.4), the four-acceleration has terms that scale with γ^2 and terms that scale with γ^4, making it an inconvenient object to work with. Geometrically though, it has a clean interpretation, which comes into view once you consider the inner product between the acceleration and velocity four-vectors:

$$\begin{aligned}\bar{\boldsymbol{a}}\cdot\bar{\boldsymbol{v}} &= \gamma^5(v)(\boldsymbol{a}\cdot\boldsymbol{v}) - \gamma^3(v)(\boldsymbol{a}\cdot\boldsymbol{v}) - \gamma^5(v)\frac{\boldsymbol{a}\cdot\boldsymbol{v}}{c^2}(\boldsymbol{v}\cdot\boldsymbol{v})\\ &= \gamma^3(v)(\boldsymbol{a}\cdot\boldsymbol{v})\left[\frac{1-v^2/c^2}{1-v^2/c^2} - 1\right]\\ &= 0.\end{aligned} \tag{15.6}$$

These four-vectors are therefore (in the four-vector sense) always perpendicular! That seems odd from a classical point of view: if you move in the x-direction, and speed up, both velocity and acceleration point in the same direction and are thus certainly not perpendicular. We do have a perpendicular case of course: circular motion (with a velocity along the circle, and acceleration inwards). Something similar happens here, if you consider the world line of a particle in a spacetime diagram (see figure 15.1). You can think of this line as a curve that's parametrized by the proper time τ; points on the curve are then given by the position four-vector at time τ. The velocity four-vector is the normalized tangent to this line (and indeed, by construction, has a fixed length c). When you're moving at constant velocity, the line is straight, but if you change your velocity (i.e., you accelerate), the line curves. The acceleration four-vector both quantifies that curvature, and points in the direction that the curve is bending[2].

Because by definition $\bar{\boldsymbol{p}} = m\bar{\boldsymbol{v}}$, and $\bar{\boldsymbol{F}} = \mathrm{d}\bar{\boldsymbol{p}}/\mathrm{d}\tau$, as long as the mass is conserved ($\mathrm{d}m/\mathrm{d}\tau = 0$), we do have $\bar{\boldsymbol{F}} = m\bar{\boldsymbol{a}}$, so Newton's second law does hold for four-vectors. This result is not nearly as useful as in classical mechanics though, since as we've seen, forces transform in unwieldly manners between inertial frames, and the acceleration can only curve the trajectory in spacetime.

To see how forces and accelerations can be used for a case where you have no choice but to use them[3], consider a particle that is under constant acceleration, due to a constant three-force acting on it in the (non-inertial!) co-moving frame of the particle (e.g. due to a rocket engine attached to the particle). From the point of view of an inertial rest frame, we have

$$\frac{\mathrm{d}\boldsymbol{p}}{\mathrm{d}t} = \boldsymbol{F} \quad\text{and}\quad \frac{\mathrm{d}\boldsymbol{F}}{\mathrm{d}t} = 0. \tag{15.7}$$

[2]There is a one-on-one correspondence between these 'world curves' and affinely parametrized curves in real space of two or more dimensions. There too, you can define a tangent vector as the derivative of the position vector, which for an affinely parametrized curve is always of unit length. The derivative of the tangent vector, known as the normal, is always perpendicular to the tangent, and points in the direction in which the curve is bending; its magnitude quantifies the curvature.

[3]I'm sure you've noted the obvious pun here.

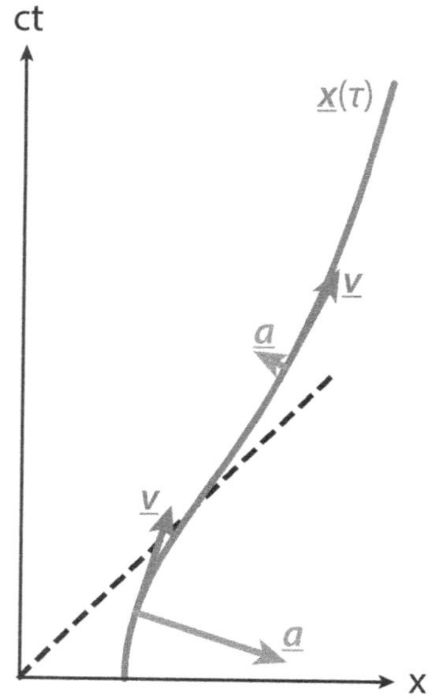

Figure 15.1: Spacetime diagram showing accelerated motion. The position four-vector gives the collection of points in spacetime that the world line passes through, parametrized by the proper time τ. The velocity four-vector (red) is the normalized tangent to that line, and the acceleration four-vector (green), which is always perpendicular to the velocity four-vector, its curvature.

Choose the x-axis to be along the direction of $\boldsymbol{F}$, and define $a = a_x = F_x/m$. Then

$$a = \frac{\mathrm{d}(p_x/m)}{\mathrm{d}t} = \frac{\mathrm{d}w_x}{\mathrm{d}t}, \tag{15.8}$$

where $\boldsymbol{w} \equiv \boldsymbol{p}/m = \gamma(v)\boldsymbol{v}$, and, as we have only motion in the positive x-direction here, we have $w_x = w$, $v_x = v$. Solving equation (15.8) for w, we get the velocity of a uniformly accelerated particle: $w(t) = w(0) + at$. Now solving for the actually measured velocity in the inertial frame (taking $w(0) = 0$), we find

$$\gamma(v(t))v(t) = w(t) = at \quad \Rightarrow \quad v^2 = a^2t^2\left(1 - \frac{v^2}{c^2}\right) \quad \Rightarrow \quad v = \frac{at}{\sqrt{1 + a^2t^2/c^2}}. \tag{15.9}$$

Figure 15.2 compares the relativistic velocity with the classical result. Unsurprisingly, they are initially identical, but once the speed starts picking up, we see that the classical results starts to deviate significantly. In particular, the relativistic result confirms that no matter how long we accelerate, our particle will never exceed the speed of light.

On a side note, we can also solve for the actual trajectory of our particle: simply integrate $\mathrm{d}x/\mathrm{d}t = v(t)$, which gives

$$x(t) = \frac{c^2}{a}\left(\sqrt{1 + \frac{a^2t^2}{c^2}} - 1\right). \tag{15.10}$$

For small values of t, we (again) recover the classical result, $x = \frac{1}{2}at^2$.

15.3 RELATIVISTIC WAVES

We've seen that in special relativity, space and time are intimately coupled. There is a classical phenomenon for which this is also the case: the waves we discussed in chapter 9. In section 9.1 we introduced the sinusoidal wave, described by (equation 9.1):

$$u(\boldsymbol{x}, t) = A\cos(\boldsymbol{k}\cdot\boldsymbol{x} - \omega t). \tag{15.11}$$

In equation (15.11) we made the wave a function of all three space coordinates, introducing a *wave vector* $\boldsymbol{k}$ rather than just the wave number k of equation (9.1). The magnitude of the wave vector is simply that of the wave number, $|\boldsymbol{k}| = k = 2\pi/\lambda$, while its direction represents the direction the (traveling) wave is moving

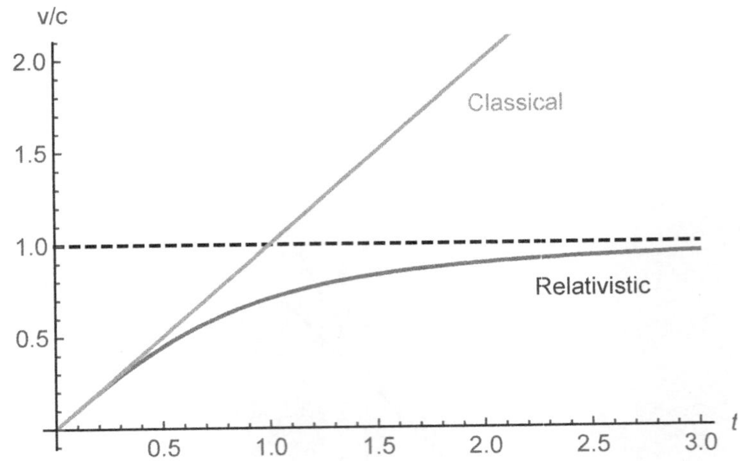

Figure 15.2: Velocity, as measured in an inertial rest frame, for a particle that undergoes constant acceleration, according to classical mechanics (orange) and special relativity (blue).

in. When written like equation (15.11), you may guess that there exists a wave four-vector combining the temporal and spatial properties of the wave, and you would be correct. If we define

$$\bar{\boldsymbol{k}} = (\omega/c, \boldsymbol{k}), \tag{15.12}$$

then the argument of the cosine in equation (15.11), i.e., the phase $\phi(\boldsymbol{x}, t)$ of the wave at the given point in space and time, is given by

$$\phi(\boldsymbol{x}, t) \equiv \boldsymbol{k}\cdot\boldsymbol{x} - \omega t = -(\bar{\boldsymbol{k}}\cdot\bar{\boldsymbol{x}}). \tag{15.13}$$

We've already shown that dot products of two four-vectors are invariant under Lorentz transformations; as ϕ is a scalar (and thus invariant) and $\bar{\boldsymbol{x}}$ a four-vector, it follows that $\bar{\boldsymbol{k}}$ is indeed also a four-vector.

The main application of relativistic waves is light itself - in its occurrence as a wave. The wave four-vector of a light beam traveling in the positive x-direction is given by

$$\bar{\boldsymbol{k}} = (k, k, 0, 0), \tag{15.14}$$

where we used that for light, $\omega = ck$ (see equation 9.2). Unsurprisingly, this looks exactly like equation (14.3) for the four-momentum of a photon - especially because that the energy of a photon is $E = hc/\lambda = hck/2\pi$. Up to a physical constant, the wave and momentum four vector of light are thus identical:

$$\bar{\boldsymbol{p}}_{\text{photon}} = \frac{h}{2\pi}\bar{\boldsymbol{k}}_{\text{photon}} = \hbar\bar{\boldsymbol{k}}_{\text{photon}}. \tag{15.15}$$

The combination $h/2\pi$ occurs so often that it got its own symbol, $\hbar$ ('h-bar'). Note that equation (15.15) holds for light only.

You might expect that there is little more to say about light. After all, the light postulate ensures that the speed of light will be the same for all observers. Yet, different observers can observe the same ray of light (or the same photon) differently: although its speed is invariant, its frequency (and thus its color, as well as its momentum) is not! To see what happens, let us start with a stationary light source emitting rays in the positive x-direction in some system S, so the wave four-vector is given by equation (15.14). We now Lorentz-transform to a system S' moving with speed u in the x direction. The wave four-vector as measured by an observer in S' is simply the Lorentz transform of (15.14):

$$\bar{\boldsymbol{k}}' = \begin{pmatrix} \gamma(u) & -\gamma(u)\frac{u}{c} & 0 & 0 \\ -\gamma(u)\frac{u}{c} & \gamma(u) & 0 & 0 \\ 0 & 0 & 1 & 0 \\ 0 & 0 & 0 & 1 \end{pmatrix}\begin{pmatrix} k \\ k \\ 0 \\ 0 \end{pmatrix} = \begin{pmatrix} \gamma(u)k(1-\frac{u}{c}) \\ \gamma(u)k(1-\frac{u}{c}) \\ 0 \\ 0 \end{pmatrix} = \begin{pmatrix} k' \\ k' \\ 0 \\ 0 \end{pmatrix}. \tag{15.16}$$

We find that the moving observer still sees the light moving in the positive x direction with speed c, but with a different wave number k', and thus a different frequency $\omega' = ck'$, given by

$$\omega' = \gamma(u)\left(1 - \frac{u}{c}\right)\omega = \sqrt{\frac{1-u/c}{1+u/c}}\,\omega. \tag{15.17}$$

Equation (15.17) gives the *relativistic Doppler effect*: a shift in observed frequency due to the motion of the observer, just as we found for sound in section 9.7. In fact, equation (15.17) reduces to (9.20) for small velocities $u << c$. In addition to the 'sound effect' where we account for the stretching or compression of the waves due to the motion of the observer, the relativistic Doppler effect also accounts for the time dilation between the two observers (it can also be derived by combining these two effects, as is done in many textbooks, see problem 15.3.2). Unlike for sound, there is also a transverse relativistic Doppler effect (entirely due to the time dilation), for which we can find the expression by replacing the ray traveling in the positive x direction in equation (15.16) with one traveling in the positive y direction.

15.4 PROBLEMS

15.1 Two spaceships are connected by a strong cable. Both ships are initially at rest, and ignite their engines at the same time. Their accelerations are identical at every point in time. Does the cable stay intact? *Hint*: use a spacetime diagram to analyze what happens.

15.2 In classical mechanics, the Doppler shift in the wavelength of an object moving towards the observer at speed u is given by (equation 9.19):

$$\lambda_{\text{obs}} = \frac{v-u}{v}\lambda,$$

where v is the speed of the wave (usually sound). To get this result, we compared what happens with the source and the wave in a fixed time interval Δt. As you now know, this result cannot hold at speeds close to that of light, because in that case there will be a significant effect due to time dilation. In this problem, we'll therefore redo the calculation to account for relativistic effects.

We consider a distant source of light that moves with velocity $\boldsymbol{u}$. At time $t = 0$ (for both the source and the stationary observer), the source emits a signal (this could be a wavecrest, but the argument holds for any signal). A time $\Delta t'$ later, as measured on the clock moving with the source, the source emits a second signal, see figure 15.3.

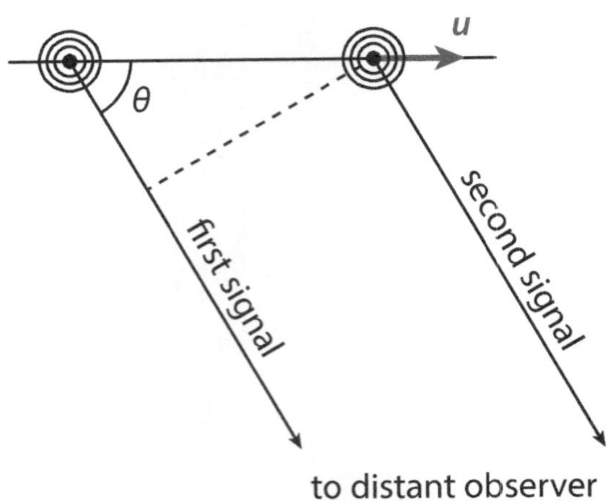

Figure 15.3: Setting for calculating the relativistic Doppler effect. As the (stationary) observer is distant, the signals traveling to the observer from the source can be taken parallel to each other; they make an angle θ with the direction in which the source moves. Note that both the distance between the position of the source at the emission of the first and second signal and the difference in distance traveled by the signal matter.

(a) Determine the time interval Δt between the emission of the first and second signal as measured on the clock of the stationary observer.

(b) Determine the change in distance Δx between the (stationary) observer and the (moving) source in the time interval between the two signals, as measured by the stationary observer.

(c) Now determine the time interval Δt_{obs} between the arrival of the first and second signal at the location of the observer.

(d) From your answers at (a-c), show that the observed frequency ν_{obs} is related to the source's frequency ν_{s} through

$$\nu_{\text{obs}} = \frac{\sqrt{1-u^2/c^2}}{1+(u/c)\cos\theta}\nu_{\text{s}}. \tag{15.18}$$

Note that equation (15.18) reduces to (15.17) in the case that the source is moving radially away from the observer.

(e) From equation (15.18), find the expression for the (relativistic) transverse Doppler shift for the case that the source is moving in a direction perpendicular to the observer's line of sight (i.e., $\theta = 90°$). How can you tell that in this case the Doppler shift is exclusively due to time dilation?

APPENDICES

A

MATH

A.1 VECTOR BASICS

Classical mechanics describes the motion of bodies as they move through space. To describe a motion in space it is not sufficient to give a position and a speed: you need a *direction* as well. Therefore we work with *vectors*: mathematical objects that have both a *magnitude* and a *direction*. If you tell me you're moving, I know something, but not much; I'll know more if you tell me you're moving at walking speed, and have full information of your velocity once you tell me that you're moving at walking speed towards the coffee machine. Although in principle we could make do with specifying a magnitude and direction of every vector in this way, it is often more convenient to express our vectors in a *basis*. To do so, we choose an (arbitrary) origin, and as many basis vectors as we have spatial dimensions, in such a way that they are not parallel to one another, and usually mutually perpendicular (orthogonal) and of unit length (orthonormal). Then we *decompose* our vector by giving its *components* along each of the basis vectors. The most common choice is to use a Cartesian basis, of two or three (depending on spatial dimension) basis vectors of unit length pointing in the standard x, y and z directions, and indicated as $\hat{\boldsymbol{x}}$, $\hat{\boldsymbol{y}}$ and $\hat{\boldsymbol{z}}$, or (rather annoyingly) sometimes as $\boldsymbol{i}$, $\boldsymbol{j}$ and $\boldsymbol{k}$, the latter especially in American textbooks. Other often encountered systems are polar coordinates (2D) and cylindrical and spherical coordinates (3D), see the mathematical appendix for more background on those. To write our vectors, we now specify the components in each direction, writing for example $\boldsymbol{v} = 3\hat{\boldsymbol{x}} + 3\hat{\boldsymbol{y}}$ for a vector (in boldface) representing a speed of $3\sqrt{2}$ and a direction making an angle of 45° with the horizontal.

Vectors can be added and subtracted just like scalars - simply add and subtract them by component. Graphically, you add two vectors by putting them head-to-tail: you can find the sum of two vectors $\boldsymbol{v}$ and $\boldsymbol{w}$ by putting the start of $\boldsymbol{w}$ at the end of $\boldsymbol{v}$, the sum $\boldsymbol{v}+\boldsymbol{w}$ then points from the start of $\boldsymbol{v}$ to the end of $\boldsymbol{w}$. You can also multiply a vector by a scalar, by multiplying every component of the vector with that scalar. Graphically, this means that you extend the length of the vector with the scalar factor you just multiplied with.

You can't take the product of two vectors like you would two scalars. There are however two vector operations that closely resemble the product, known as the inner (or dot) and outer (or cross) product, see figure A.1. The dot product represents the length of the projection of one vector on another (and thus gives a scalar); it is zero for perpendicular vectors, and the dot product of a vector with itself gives the square of its length. To calculate the dot product of two vectors, sum the products of their components: if $\boldsymbol{v} = v_x\hat{\boldsymbol{x}} + v_y\hat{\boldsymbol{y}}$ and $\boldsymbol{w} = w_x\hat{\boldsymbol{x}} + w_y\hat{\boldsymbol{y}}$, then $\boldsymbol{v}\cdot\boldsymbol{w} = v_x w_x + v_y w_y$. You can use the dot product to find the angle between two vectors, using standard geometry, which gives

$$\cos\theta = \frac{\boldsymbol{v}\cdot\boldsymbol{w}}{|\boldsymbol{v}||\boldsymbol{w}|} = \frac{v_x w_x + v_y w_y}{|\boldsymbol{v}||\boldsymbol{w}|}, \tag{A.1}$$

where $|\boldsymbol{v}|$ and $|\boldsymbol{w}|$ are the lengths of vectors $\boldsymbol{v}$ and $\boldsymbol{w}$, respectively. The cross product is only defined for three-dimensional vectors, say $\boldsymbol{v} = v_x\hat{\boldsymbol{x}} + v_y\hat{\boldsymbol{y}} + v_z\hat{\boldsymbol{z}}$ and $\boldsymbol{w} = w_x\hat{\boldsymbol{x}} + w_y\hat{\boldsymbol{y}} + w_z\hat{\boldsymbol{z}}$. The result is another vector, with a direction perpendicular to the plane spanned by $\boldsymbol{v}$ and $\boldsymbol{w}$, and a magnitude equal to the area of the parallelogram bounded by them. The cross product is most easily expressed in column vector form:

$$\boldsymbol{v}\times\boldsymbol{w} = \begin{pmatrix} v_x \\ v_y \\ v_z \end{pmatrix} \times \begin{pmatrix} w_x \\ w_y \\ w_z \end{pmatrix} = \begin{pmatrix} v_y w_z - v_z w_y \\ v_z w_x - v_x w_z \\ v_x w_y - v_y w_x \end{pmatrix}. \tag{A.2}$$

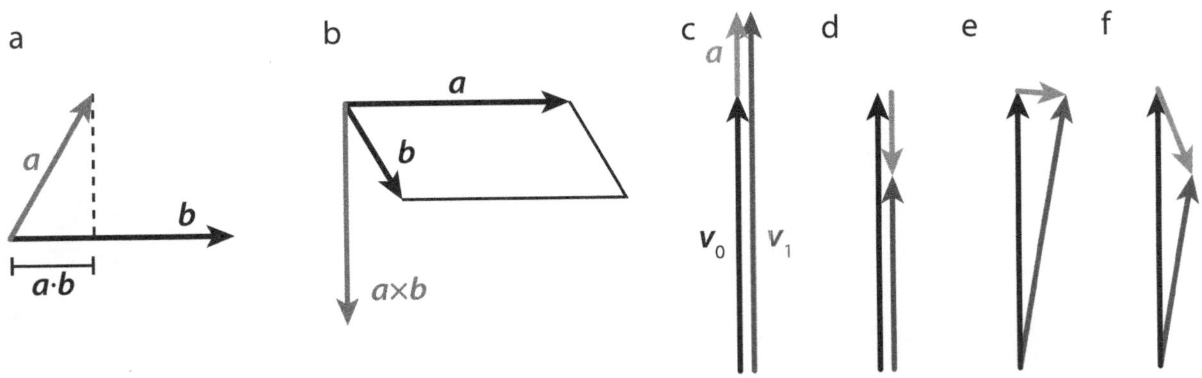

Figure A.1: Products and derivatives of vectors. (a) The dot product of two vectors $\boldsymbol{a}$ and $\boldsymbol{b}$ gives the length of the projection of $\boldsymbol{a}$ on $\boldsymbol{b}$ (or vice versa). (b) The cross product of two vectors $\boldsymbol{a}$ and $\boldsymbol{b}$ gives a vector perpendicular to the plane spanned by $\boldsymbol{a}$ and $\boldsymbol{b}$ and with a magnitude equal to the area of the parallelogram spanned by $\boldsymbol{a}$ and $\boldsymbol{b}$. (c-f) Acceleration as a change in velocity. In every panel, the original velocity is shown in black, the velocity a short time Δt later is shown in blue, and the acceleration is shown in red. (c) Acceleration for an increase in the magnitude of the velocity. (d) Acceleration for a decrease in the magnitude of the velocity. (e) Acceleration for a change in direction of the velocity at constant magnitude. (f) Acceleration for a change in velocity that involves both a change in direction and a decrease in magnitude.

The cross product of a vector with itself is zero.

Vectors can be *functions*, just like scalar quantities: they can depend on one or more parameters, like position or time. Also, again just like scalar functions, you can calculate a *rate of change* of vector function as you move through parameter values, for instance asking how the velocity of a car changes as a function of time. An instantaneous rate of change is simply a derivative, which is calculated in exactly the same manner as the derivative of a scalar function. For example, the rate of change of the velocity, known as the *acceleration* $\boldsymbol{a}$, is defined as:

$$\boldsymbol{a} = \lim_{\Delta t \to 0} \frac{\boldsymbol{v}(t+\Delta t) - \boldsymbol{v}}{\Delta t}. \tag{A.3}$$

Since the velocity itself is the derivative of the *position* $\boldsymbol{x}(t)$, the acceleration is also the second derivative of the position. Time derivatives occur so frequently in classical mechanics that we use a special notation for them: a first derivative is indicated by a dot on top of the quantity, and a second derivative by a double dot - so we have $\boldsymbol{a} = \dot{\boldsymbol{v}} = \ddot{\boldsymbol{x}}$.

Vector derivatives are somewhat richer than those of scalar functions, since there are more ways that a vector can change. Like a scalar function, the magnitude of a vector can increase or decrease. Moreover, its direction can also change, which also means that it has a nonzero derivative, and of course, you can have a combination of a change in magnitude and a change in direction, see figure A.1.

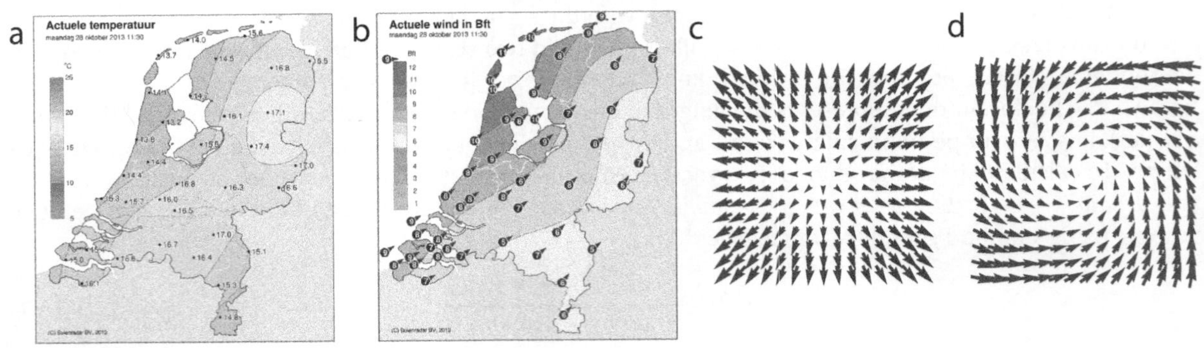

Figure A.2: Examples of scalar and vector fields. (a) Temperature (a scalar) at every point in the Netherlands at 11:30 on October 28, 2013. Lines of equal temperature are drawn in black; the gradient of the temperature field is everywhere perpendicular to these lines. (b) Wind velocity (a vector) at every point in the Netherlands at the same time as the temperature map. Colors correspond to the magnitude of the velocity. (a and b) from [30]. (c) Example of a vector field with a large divergence and zero curl. (d) Example of a vector field with a large curl (and low but nonzero divergence).

Functions (scalar or vector) that are defined at every point in space are sometimes called *fields*. Examples are the temperature (scalar) and wind (vector) at every point on the planet, see figure A.2. Just like you can calculate the rate of change of a function in time, you can also consider how a function changes in space. For a scalar function, this quantity is a vector, known as the *gradient*, defined as the vector of partial derivatives. For a function $f(x,y,z)$, we have:

$$\nabla f = \frac{\partial f}{\partial x}\hat{\boldsymbol{x}} + \frac{\partial f}{\partial y}\hat{\boldsymbol{y}} + \frac{\partial f}{\partial z}\hat{\boldsymbol{z}} = \begin{pmatrix} \partial f/\partial x \\ \partial f/\partial y \\ \partial f/\partial z \end{pmatrix}. \tag{A.4}$$

The direction of ∇f is the direction of maximal change, and its magnitude tells you how quickly the function changes in that direction. For a vector field $\boldsymbol{v}$, we can't take the gradient, but we can use the 'vector' ∇ of partial derivatives combined with either the dot or cross product. The first option is known as the *divergence* of $\boldsymbol{v}$, and tells you how quickly $\boldsymbol{v}$ spreads out; the second is the *curl* of $\boldsymbol{v}$ and tells you how much $\boldsymbol{v}$ rotates:

$$\mathrm{div}(\boldsymbol{v}) = \nabla \cdot \boldsymbol{v} = \frac{\partial v_x}{\partial x} + \frac{\partial v_y}{\partial y} + \frac{\partial v_z}{\partial z}, \tag{A.5}$$

$$\mathrm{curl}(\boldsymbol{v}) = \nabla \times \boldsymbol{v} = \begin{pmatrix} \partial_y v_z - \partial_z v_y \\ \partial_z v_x - \partial_x v_z \\ \partial_x v_y - \partial_y v_x \end{pmatrix}, \tag{A.6}$$

where $\partial_x = \partial/\partial x$, and so on.

A.2 POLAR COORDINATES

You can specify any point in the plane by specifying its projection on two perpendicular axes - we typically call these the x and y-axes and x and y coordinates. In this Cartesian system (named after Descartes), we identify *unit vectors* $\hat{\boldsymbol{x}}$ and $\hat{\boldsymbol{y}}$, pointing along their respective axes, and being of unit length. A position $\boldsymbol{r}$ can then be decomposed in the two directions: $\boldsymbol{r} = r_x\hat{\boldsymbol{x}} + r_y\hat{\boldsymbol{y}}$, with $r_x = \boldsymbol{r}\cdot\hat{\boldsymbol{x}}$ and $r_y = \boldsymbol{r}\cdot\hat{\boldsymbol{y}}$. Alternatively, we can write $\hat{\boldsymbol{x}} = \begin{pmatrix} 1 \\ 0 \end{pmatrix}$ and $\hat{\boldsymbol{y}} = \begin{pmatrix} 0 \\ 1 \end{pmatrix}$, which gives for $\boldsymbol{r}$:

$$\boldsymbol{r} = r_x\hat{\boldsymbol{x}} + r_y\hat{\boldsymbol{y}} = \begin{pmatrix} r_x \\ r_y \end{pmatrix}.$$

Instead of specifying the x and y coordinates of our position, we could also uniquely identify it by giving two different numbers: its distance to the origin r, and the angle θ the line to the origin makes with a fixed reference axis (typically the x-axis), see figure A.3. Invoking the Pythagorean theorem and basic trigonometry, we readily find $r = \sqrt{r_x^2 + r_y^2}$ and $\tan\theta = r_y/r_x$. We call r the length of the vector $\boldsymbol{r}$. We could also invert the relations for r and θ so we can get the Cartesian components if the length and angle are known: $r_x = r\cos\theta$ and $r_y = r\sin\theta$.

Like the Cartesian basis vectors $\hat{\boldsymbol{x}}$ and $\hat{\boldsymbol{y}}$, which point in the direction of increasing x and y values, we can also define unit vectors pointing in the direction of increasing r and θ. These directions do depend on our position in space, but they do have a clear geometrical interpretation: $\hat{\boldsymbol{r}}$ always points radially outward from the origin, and $\hat{\boldsymbol{\theta}}$ in the direction you'd move if you'd be making a counterclockwise rotation about the origin. Given a position vector $\boldsymbol{r}$, finding the vector in the direction of increasing r is very easy: $\hat{\boldsymbol{r}} = \boldsymbol{r}/r$. The expression for $\boldsymbol{r}$ in our new polar basis $(\hat{\boldsymbol{r}},\hat{\boldsymbol{\theta}})$ is almost tautological: $\boldsymbol{r} = r\hat{\boldsymbol{r}}$.

Relating the polar basis vectors to the Cartesian ones is straightforward. We have:

$$\boldsymbol{r} = r_x\hat{\boldsymbol{x}} + r_y\hat{\boldsymbol{y}} = r\hat{\boldsymbol{r}},$$

and using $r_x = r\cos\theta$, $r_y = r\sin\theta$ we also have

$$\boldsymbol{r} = r\cos\theta\hat{\boldsymbol{x}} + r\sin\theta\hat{\boldsymbol{y}}.$$

We thus find that $\hat{\boldsymbol{r}} = \cos\theta\hat{\boldsymbol{x}} + \sin\theta\hat{\boldsymbol{y}}$. For $\hat{\boldsymbol{\theta}}$ we note that to rotate around the origin, the direction of motion needs to be perpendicular to $\hat{\boldsymbol{r}}$. There are of course two such directions - we pick the sign by demanding that

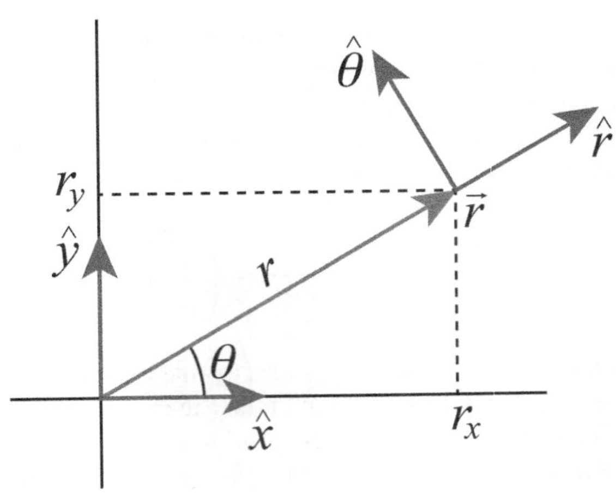

Figure A.3: Polar coordinate system and polar unit vectors. A position $\boldsymbol{r}$ in the plane can be specified either by giving its projections on two reference axes (along the $\hat{\boldsymbol{x}}$ and $\hat{\boldsymbol{y}}$ direction), or by giving its distance r to the origin, and the angle θ the line to the origin makes with the x axis. The polar basis vectors are defined as pointing in the direction of increasing r (i.e., radially outward), and increasing θ (i.e., rotating counterclockwise around the origin).

the counterclockwise rotation is positive. This gives $\hat{\boldsymbol{\theta}} = (r_y/r)\hat{\boldsymbol{x}} - (r_x/r)\hat{\boldsymbol{y}} = \sin\theta\hat{\boldsymbol{x}} - \cos\theta\hat{\boldsymbol{y}}$. Written out as vectors, we have:

$$\hat{\boldsymbol{r}} = \begin{pmatrix} \cos\theta \\ \sin\theta \end{pmatrix}, \qquad \hat{\boldsymbol{\theta}} = \begin{pmatrix} \sin\theta \\ -\cos\theta \end{pmatrix}. \tag{A.7}$$

Note that

$$\hat{\boldsymbol{r}} = \frac{\partial\hat{\boldsymbol{\theta}}}{\partial\theta}, \qquad \hat{\boldsymbol{\theta}} = -\frac{\partial\hat{\boldsymbol{r}}}{\partial\theta}. \tag{A.8}$$

Naturally, we can also express the Cartesian basis in terms of the polar ones:

$$\hat{\boldsymbol{x}} = \cos\theta\hat{\boldsymbol{r}} + \sin\theta\hat{\boldsymbol{\theta}}, \qquad \hat{\boldsymbol{y}} = \sin\theta\hat{\boldsymbol{r}} - \cos\theta\hat{\boldsymbol{\theta}}. \tag{A.9}$$

A.3 SOLVING DIFFERENTIAL EQUATIONS

A *differential equation* is an equation which contains derivatives of the function to be determined. They can be very simple. For example, you may be given the (constant) velocity of a car, which is the derivative of its position, which we'd write mathematically as:

$$v = \frac{\mathrm{d}x}{\mathrm{d}t} = v_0. \tag{A.10}$$

To determine where the car ends up after one hour, we need to solve this differential equation. We also need a second piece of information: where the car was at some reference time (usually $t = 0$), the *initial condition*. If $x(0) = 0$, you don't need advanced maths skills to figure out that $x(1\text{ hour}) = v_0 \cdot (1\text{ hour})$. Unfortunately, things aren't usually this easy.

Before we proceed to a few techniques for solving differential equations, we need some terminology. The *order* of a differential equation is the order of the highest derivative found in the equation; equation (A.10) is thus of first order. A differential equation is called *ordinary* if it only contains derivatives with respect to one variable, and *partial* if it contains derivatives to multiple variables. The equation is *linear* if it does not contain any products of (derivatives of) the unknown function. Finally, a differential equation is *homogeneous* if it only contains terms that contain the unknown function, and *inhomogeneous* if it also contains other terms. Equation (A.10) is ordinary and inhomogeneous, as the v_0 term on the right does not contain the unknown function $x(t)$. In the sections below, we discuss the various cases you'll encounter in this book; there are many others (many of which can't be solved explicitly) to which a whole subfield of mathematics is dedicated.

A.3.1 FIRST-ORDER LINEAR ORDINARY DIFFERENTIAL EQUATIONS

Suppose we have a general equation of the form

$$a(t)\frac{\mathrm{d}x}{\mathrm{d}t} + b(t)x(t) = f(t), \tag{A.11}$$

where $a(t)$, $b(t)$ and $f(t)$ are known functions of t, and $x(t)$ is our unknown function. Equation (A.11) is a first-order, ordinary, linear, inhomogeneous differential equation. In order to solve it we will use two techniques that are tremendously useful: separation of variables and separation into homogeneous and particular solutions.

Suppose we had $f(t) = 0$. Then, if we had two solutions $x_1(t)$ and $x_2(t)$ of equation (A.11), we could construct a third as $x_1(t) + x_2(t)$ (or any linear combination of $x_1(t)$ and $x_2(t)$), since the equation is linear. Now since $f(t)$ is not zero, we can't do this, but we can do something else. First, we find the most general solution to the equation where $f(t) = 0$, which we call the *homogeneous solution* $x_{\mathrm{h}}(t)$. Second, we find a solution (any at all) of the full equation (A.11), which we call the *particular solution* $x_{\mathrm{p}}(t)$. The full solution is then the sum of these two solutions, $x(t) = x_{\mathrm{h}}(t) + x_{\mathrm{p}}(t)$. You may worry that there may be multiple particular solutions: how would we pick the 'right' one? Fortunately, we don't need to worry: the homogeneous solution will contain an unknown variable, which will be set by the initial condition. Changing the particular solution will change the value of the variable, such that the final solution will be the same and satisfy both the differential equation and the initial condition.

To find the solution to the homogeneous equation

$$a(t)\frac{\mathrm{d}x_{\mathrm{h}}}{\mathrm{d}t} + b(t)x_{\mathrm{h}}(t) = 0, \tag{A.12}$$

we're going to use a technique called *separation of variables*. There are two variables in this system: the independent parameter t and the dependent parameter x. The trick is to get everything depending on t on one side of the equals sign, and everything depending on x on the other. To do so, we're going to treat $\mathrm{d}x/\mathrm{d}t$ as if it were an actual fraction[1]. In that case, it's not hard to see that we can re-arrange equation (A.12) to

$$\frac{1}{x_{\mathrm{h}}}\,\mathrm{d}x_{\mathrm{h}} = -\frac{b(t)}{a(t)}\,\mathrm{d}t. \tag{A.13}$$

By itself, equation (A.13) means little, but if we integrate both sides, we get something that makes sense:

$$\int \frac{1}{x_{\mathrm{h}}}\,\mathrm{d}x_{\mathrm{h}} = \log(x_{\mathrm{h}}) + C = -\int \frac{b(t)}{a(t)}\,\mathrm{d}t \tag{A.14}$$

or

$$x_{\mathrm{h}}(t) = A\exp\left[-\int \frac{b(t)}{a(t)}\,\mathrm{d}t\right], \tag{A.15}$$

where $A = \exp(C)$ is an integration constant (the unknown constant that will be set by our initial condition). Of course, in principle it may not be possible to evaluate the integral in equation (A.15), but even then the solution is valid. In practice, you'll often encounter situations in which $a(t)$ and $b(t)$ are simple functions or even constants, and the evaluation of the integral is straightforward.

Now that we have our homogeneous solution, we still need a particular one. Sometimes you're lucky, and you can easily guess one - for instance one in which $x_{\mathrm{p}}(t)$ doesn't depend on t at all. In case you're not lucky, there's are two other techniques you may try, either using *variation of constants* or finding an *integrating factor*. To demonstrate variation of constants, we'll pick a specific example, to not get lost in a bunch of abstract functions. Let $a(t) = a$ be a constant and $b(t) = bt$ be linear. The homogeneous solution then becomes $x_{\mathrm{h}}(t) = A\exp\left[-\frac{1}{2}\frac{b}{a}t^2\right]$. The constant we're going to vary is our integration constant A, so our guess for the particular solution will be

$$x_{\mathrm{p}}(t) = A(t)\exp\left[-\frac{1}{2}\frac{b}{a}t^2\right]. \tag{A.16}$$

We substitute (A.16) back into the full differential equation (A.11), which gives:

$$\left[a\frac{\mathrm{d}A}{\mathrm{d}t} - aA(t)\frac{bt}{a} + btA(t)\right]\exp\left[-\frac{1}{2}\frac{b}{a}t^2\right] = a\frac{\mathrm{d}A}{\mathrm{d}t}\exp\left[-\frac{1}{2}\frac{b}{a}t^2\right] = f(t). \tag{A.17}$$

[1] If you're worried about the validity of this shortcut, consider that we could also have left $\mathrm{d}x_{\mathrm{h}}/\mathrm{d}t$ intact, to give $\frac{1}{x_{\mathrm{h}}}\mathrm{d}x_{\mathrm{h}}/\mathrm{d}t = -b(t)/a(t)$. Integrating both sides over time gives

$$\int \frac{1}{x_{\mathrm{h}}}\,\mathrm{d}x_{\mathrm{h}}/\mathrm{d}t\,\mathrm{d}t = \int \frac{1}{x_{\mathrm{h}}}\,\mathrm{d}x_{\mathrm{h}} = -\int \frac{b(t)}{a(t)}\,\mathrm{d}t,$$

which takes us back to equation (A.14).

A big part of the left-hand side thus cancels, and that's not a coincidence - that's because it is based on the homogeneous equation. What remains is a differential equation in $A(t)$ that can be trivially solved by direct integration:

$$A(t) = \int \frac{\mathrm{d}A}{\mathrm{d}t}\mathrm{d}t = \frac{1}{a}\int f(t)\exp\left[\frac{1}{2}\frac{b}{a}t^2\right]\mathrm{d}t. \tag{A.18}$$

Again, it may not be possible to evaluate the integral in equation (A.18), but in principle the solution could be inserted in equation (A.16) to give us our particular solution, and the whole differential equation will be solved.

Alternatively, we may try to find an *integration factor* for equation (A.11). This means that we try to rewrite the left hand side of the equation as a total derivative, after which we can simply integrate to get the solution. To do so, we first divide the whole equation by $a(t)$, then look for a function $\mu(t)$ that satisfies the condition that

$$\frac{\mathrm{d}}{\mathrm{d}t}\left[\mu(t)x(t)\right] = \mu(t)\frac{\mathrm{d}x}{\mathrm{d}t} + x(t)\frac{\mathrm{d}\mu}{\mathrm{d}t} = \mu(t)\frac{\mathrm{d}x}{\mathrm{d}t} + \mu(t)\frac{b(t)}{a(t)}x(t), \tag{A.19}$$

from which we can read off that we need to solve the homogeneous equation

$$\frac{\mathrm{d}\mu}{\mathrm{d}t} = \frac{b(t)}{a(t)}\mu(t). \tag{A.20}$$

We can solve (A.20) by separation of constants, which gives us

$$\mu(t) = \exp\left(\int \frac{b(t)}{a(t)}\mathrm{d}t\right), \tag{A.21}$$

where we set the integration constant to one, as it drops out of the equation for $x(t)$ anyway. With this function $\mu(t)$, we can rewrite equation (A.11) as

$$\frac{\mathrm{d}}{\mathrm{d}t}\left[\mu(t)x(t)\right] = \mu(t)\frac{f(t)}{a(t)}, \tag{A.22}$$

which we can integrate to find $x(t)$:

$$x(t) = \frac{1}{\mu(t)}\int \mu(t)\frac{f(t)}{a(t)}\mathrm{d}t. \tag{A.23}$$

A.3.2 SECOND-ORDER LINEAR ORDINARY DIFFERENTIAL EQUATIONS WITH CONSTANT COEFFICIENTS

Second order ordinary differential equations are essential for the study of mechanics, as its central equation, Newton's second law of motion (equation 2.4) is of this type. In the case that the equation is also linear, we have some hopes of solving it analytically. There are several examples of this type of equation in the main text, especially in section 2.6, where we solve the equation of motion resulting from Newton's second law for three special cases, and section 8.1, where we study a number of variants of the harmonic oscillator.

For the case that the equation is homogeneous and has constant coefficients, we can write down the general solution[2]. The equation to be solved is of the form

$$a\frac{\mathrm{d}^2x}{\mathrm{d}t^2} + b\frac{\mathrm{d}x}{\mathrm{d}t} + cx(t) = 0. \tag{A.24}$$

For the case that $a = 0$, we retrieve a first-order differential equation, whose solution is an exponential (as can be found by separation of variables and integration): $x(t) = C\exp(ct/b)$. In many cases an exponential is also a solution of equation (A.24). To figure out which exponential, let's start with the trial function (or 'Ansatz') $x(t) = \exp(\lambda t)$, where λ is an unknown parameter. Substituting this Ansatz into equation (A.24) yields the *characteristic polynomial* for this ode:

$$a\lambda^2 + b\lambda + c = 0, \tag{A.25}$$

which almost always has two solutions:

$$\lambda_{\pm} = -\frac{b}{2a} \pm \frac{\sqrt{b^2 - 4ac}}{2a}. \tag{A.26}$$

[2]This approach generalizes to higher-order differential equations, though the characteristic polynomial may not be as easy to solve in the general case.

Note that the solutions can be real or complex. If there are two of them, we can write the general solution[3] of equation (A.24) as a linear combination of the Ansatz with the two cases:

$$x(t) = Ae^{\lambda_+ t} + Be^{\lambda_- t}, \tag{A.27}$$

where A and B are set by either initial or boundary conditions. Since the $\lambda_\pm$ may be complex, so may A and B; it's their combination that should give a real number (as $x(t)$ is real), see problem A.3.1a.

In the case that equation (A.26) gives only one solution, the corresponding exponential function is still a solution of equation (A.24), but it is not the most general one, as we only can put a single undetermined constant in front of it. We therefore need a second, independent solution. To guess one, here's a third useful trick[4]: take the derivative of our known solution, $e^{\lambda t}$, with respect to the parameter λ. This gives a second Ansatz: $te^{\lambda t}$, where $\lambda = -b/2a$. Substituting this Ansatz into equation (A.24) for the case that $c = b^2/2a$, we find:

$$\frac{\mathrm{d}^2x}{\mathrm{d}t^2} + b\frac{\mathrm{d}x}{\mathrm{d}t} + \frac{b^2}{2a}x(t) = a\left(-\frac{b}{a} + \frac{b^2}{4a^2}t\right)e^{-\frac{bt}{2a}} + b\left(1 - \frac{b}{2a}t\right)e^{-\frac{bt}{2a}} + \frac{b^2}{4a}te^{-\frac{bt}{2a}} = 0, \tag{A.28}$$

so our Ansatz is again a solution. For this special case, the general solution is therefore given by

$$x(t) = Ae^{-\frac{bt}{2a}} + Bte^{-\frac{bt}{2a}}. \tag{A.29}$$

In section 8.2, where we discuss the damped harmonic oscillator, the special case corresponds to the critically damped oscillator. We get an underdamped oscillator when the roots of the characteristic polynomial are complex, and an overdamped one when they are real.

A.3.3 SECOND-ORDER LINEAR ORDINARY DIFFERENTIAL EQUATIONS OF EULER TYPE

There's a second class of linear ordinary differential equations that we can solve explicitly: those of Euler (or Cauchy-Euler) type, where the coefficient in front of a derivative contains the variable to the power of the derivative, i.e., for a second-order differential equation, we have as the most general form:

$$ax^2\frac{\mathrm{d}^2y}{\mathrm{d}x^2} + bx\frac{\mathrm{d}y}{\mathrm{d}x} + cy(x) = 0. \tag{A.30}$$

Note that we are now solving for $y(x)$; we do so because this type of equation typically occurs in the context of position- rather than time-dependent functions. An example is the Laplace equation ($\nabla^2 y = 0$) in polar coordinates. Like for the second order ode with constant coefficients, the ode of Euler type can be generalized to higher-order equations.

There are (at least) two ways to solve equation (A.30): through an Ansatz, and through a change of variables. For the Ansatz, note that for any polynomial, the derivative of each term reduces the power by one, and here we're multiplying each such term with the variable to the power the number of derivatives[5]. This suggests we simply try a polynomial, so our Ansatz here will be $y(x) = x^n$. Substituting in equation (A.30) gives:

$$ax^2n(n-1)x^{n-2} + bxnx^{n-1} + cx^n = [an(n-1) + bn + c]x^n = 0, \tag{A.31}$$

so we get another second order polynomial to solve, this time in n:

$$an^2 + (b-a)n + c = 0 \quad \Rightarrow \quad n_\pm = \frac{1}{2} - \frac{b}{2a} \pm \frac{1}{2}\sqrt{(a-b)^2 - 4ac}. \tag{A.32}$$

If the roots in equation (A.32) are both real (the most common case in physics problems), we have two independent solutions, and we are done. If the roots are complex, we also have two independent solutions,

[3] You may wonder how we know that this is the most general solution of equation (A.24). The argument follows from the proof of the Picard Lindelöf theorem, which states that under fairly weak conditions on the ode (which are met by equation (A.24), there exists a single unique solution, which depends on as many free parameters as the order of the equation. For example, if you know the position and its derivative at $t = 0$, everything else is fixed: equation (A.24) gives you the value of the second derivative, from which you can calculate the value of the derivative and the function itself at some later time Δt, and so on. As equation (A.27) has two unknowns, we've used up all our degrees of freedom in determining them, and as the solution is unique, it cannot contain another term.

[4] Alternatively, you could use a variant of variation of constants, known as reduction of order; see section A.3.4 and problem A.3.2.

[5] An admittedly complicated sentence, but if you find yourself wondering about it, just press on, and return later, when it will hopefully be obvious.

though they involve complex powers of x; like for the equation with constant coefficients, we can rewrite these as real functions with Euler's formula (see problem A.3.1b). For the case that we have only one root, we again apply our trick to get a second: we try $\mathrm{d}x^n/\mathrm{d}n = x^n \ln(x)$, which turns out to be indeed a solution (problem A.3.1c), and the general solution is again a linear combination of the two solutions found.

Alternatively, we could have solved equation (A.30) by a change of variables. Although this method is occasionally useful (and so it's good to be aware of its existence), there is no systematic way of deriving which change of variables will do the trick, so you'll have to go by trial-and-error (without a priori guarantee of success). In this case, this process leads to the following substitution:

$$x = e^t, \quad y(x) = y(e^t) \equiv \phi(t), \tag{A.33}$$

where we introduce $\phi(t)$ for convenience. Taking derivatives of $y(x)$ with the chain rule gives

$$\frac{\mathrm{d}y}{\mathrm{d}x} = \frac{\mathrm{d}y}{\mathrm{d}t}\frac{\mathrm{d}t}{\mathrm{d}x} = \frac{1}{x}\frac{\mathrm{d}\phi}{\mathrm{d}t}, \quad \frac{\mathrm{d}^2 y}{\mathrm{d}x^2} = \frac{1}{x^2}\left(\frac{\mathrm{d}^2\phi}{\mathrm{d}t^2} - \frac{\mathrm{d}\phi}{\mathrm{d}t}\right), \tag{A.34}$$

where we kept the $1/x$ and $1/x^2$ as they'll cancel the coefficients in equation (A.30). Substituting (A.33) and (A.34) in equation (A.30), we get

$$a\frac{\mathrm{d}^2\phi}{\mathrm{d}t^2} + (b-a)\frac{\mathrm{d}\phi}{\mathrm{d}t} + c\phi(t) = 0, \tag{A.35}$$

which is a second order differential equation with constant coefficients, and thus of the form given in equation (A.24). We therefore know how to find its solutions, and can use equation (A.33) to transform those solutions back to functions $y(x)$.

A.3.4 Reduction of order

If you find yourself with a non-homogeneous second order differential equation where the homogeneous equation either has constant coefficients or is of Euler type, you can again use the technique of variation of constants to find a particular solution. A similar technique, known as reduction of order, may help you find solutions to a second (or higher) order equation where the coefficients are not constant. In order to be able to use this technique, you need to know a solution to the homogeneous equation, so it is not as universally applicable as the techniques in the previous two sections, but still frequently very helpful.

Let us write the general non-homogeneous second-order linear differential equation as

$$\frac{\mathrm{d}^2 y}{\mathrm{d}x^2} + p(t)\frac{\mathrm{d}y}{\mathrm{d}x} + q(x)y(x) = r(x). \tag{A.36}$$

Note that this is the most general form: if there is a coefficient (constant or otherwise) in front of the second derivative, we simply divide the whole equation by that coefficient and redefine the coefficients to match equation (A.36). Now suppose we have a solution $y_1(x)$ of the homogeneous equation (so for the case that $r(x) = 0$). As the equation is homogeneous, for any constant v the function $vy_1(x)$ will also be a solution. As an Ansatz for the second solution, we'll try a variant of variation of constants, and take

$$y_2(x) = v(x)y_1(x), \tag{A.37}$$

where $v(x)$ is an arbitrary function. Substituting (A.37) back into (A.36), we find

$$y_1(x)\frac{\mathrm{d}^2 v}{\mathrm{d}x^2} + \left[2\frac{\mathrm{d}y_1}{\mathrm{d}x} + p(x)y_1(x)\right]\frac{\mathrm{d}v}{\mathrm{d}x} + \left[\frac{\mathrm{d}^2 y_1}{\mathrm{d}x^2} + p(x)\frac{\mathrm{d}y_1}{\mathrm{d}x} + q(x)y_1(x)\right]v(x) = r(x). \tag{A.38}$$

We recognize the prefactor of $v(x)$ as exactly the homogeneous equation, which $y_1(x)$ satisfies, so this term vanishes. Now defining $w(x) = \mathrm{d}v/\mathrm{d}x$, we are left with a first-order equation for $w(x)$:

$$y_1(x)\frac{\mathrm{d}w}{\mathrm{d}x} + \left[2\frac{\mathrm{d}y_1}{\mathrm{d}x} + p(x)y_1(x)\right]w(x) = r(x). \tag{A.39}$$

Equation (A.39) is a first-order linear differential equation, and can be solved by the techniques from section (A.3.1). Integrating the equation $w(x) = \mathrm{d}v/\mathrm{d}x$ then gives us $v(x)$, and hence the second solution (A.37) of the (inhomogeneous) second order differential equation.

A.3.5 POWER SERIES SOLUTIONS

If none of the techniques in the sections above apply to your differential equation, there's one last Ansatz you can try: a power series expansion of your solution. To illustrate, we'll again pick a concrete example: Legendre's differential equation, given by

$$\frac{\mathrm{d}}{\mathrm{d}x}\left[(1-x^2)\frac{\mathrm{d}y}{\mathrm{d}x}\right]+n(n+1)y(x)=(1-x^2)\frac{\mathrm{d}^2y}{\mathrm{d}x^2}-2x\frac{\mathrm{d}y}{\mathrm{d}x}+n(n+1)y(x)=0, \tag{A.40}$$

where n is an integer. As an Ansatz for the solution, we'll try a power series expansion of $y(x)$:

$$y(x)=\sum_{k=0}^{\infty}a_k x^k. \tag{A.41}$$

Our task is now to find numbers a_k (many of which may be zero) such that (A.41) is a solution of (A.40). Fortunately, we can simply substitute our trial solution and re-arrange to get

$$\begin{aligned}
0 &= (1-x^2)\frac{\mathrm{d}^2}{\mathrm{d}x^2}\left(\sum_{k=0}^{\infty}a_k x^k\right)-2x\frac{\mathrm{d}}{\mathrm{d}x}\left(\sum_{k=0}^{\infty}a_k x^k\right)+n(n+1)\sum_{k=1}^{\infty}a_k x^k && \text{(A.42a)}\\
&= (1-x^2)\left(\sum_{k=0}^{\infty}k(k-1)a_k x^{k-2}\right)-2x\left(\sum_{k=0}^{\infty}ka_k x^{k-1}\right)+n(n+1)\sum_{k=1}^{\infty}a_k x^k && \text{(A.42b)}\\
&= \sum_{k=0}^{\infty}\left[(-k(k-1)-2k+n(n+1))\,a_k x^k+k(k-1)a_k x^{k-2}\right] && \text{(A.42c)}\\
&= \sum_{k=0}^{\infty}\left[(-k(k+1)+n(n+1))\,a_k+(k+2)(k+1)a_{k+2}\right]x^k, && \text{(A.42d)}
\end{aligned}$$

where in the last line, we 'shifted' the index of the last term[6]. We do so in order to get at an expression for the coefficient of x^i for any value of k. As the functions x^k are linearly independent[7] (i.e., you can't write x^k as a linear combination of other functions x^m where $m \neq k$), the coefficient of each of the powers in the sum in equation (A.42d) has to vanish for the sum to be identically zero. This gives us a *recurrence relation* between the coefficients a_k:

$$a_{k+2}=\frac{k(k+1)-n(n+1)}{(k+2)(k+1)}a_k. \tag{A.43}$$

Given the values of a_0 and a_1 (the two degrees of freedom that our second-order differential equation allows us), we can repeatedly apply equation (A.43) to get all coefficients. Note that for $k=n$ the coefficient equals zero. Therefore, if for an even value of n, we set $a_1=0$, and for an odd value of n, we set $a_0=0$, we get a finite number of nonzero coefficients. The resulting solutions are polynomials, characterized by the number n; in this case, they're known as the *Legendre polynomials*, typically denoted $P_n(x)$, and normalized (by setting the value of the remaining free coefficient) such that $P_n(1)=1$. Table A.1 lists the first five, which are also plotted in figure A.4a.

Legendre polynomials have many other interesting properties (many of which can be found in either math textbooks or on their Wikipedia page). They occur frequently in physics, for example in solving problems involving Newtonian gravity or Laplace's equation from electrostatics.

If we replace the $n(n+1)$ factor in the Legendre differential equation with an arbitrary number λ, the series solution remains a solution, but it no longer terminates[8]. There are many other differential equations

[6]Note that we can do this because the index k is a 'dummy', i.e., an index that we sum over and thus does not appear in the end answer. We could just as well have called it j, or we could define $j=k-1$, and re-write the sum over k as

$$\sum_{k=0}^{\infty}a_k x^k=\sum_{j=-1}^{\infty}a_{j+1}x^{j+1}=\sum_{j=-1}^{\infty}a_{j+1}x^{j+1}.$$

Therefore

$$\sum_{k=0}^{\infty}ka_k x^{k-1}=\sum_{j=-1}^{\infty}(j+1)a_{j+1}x^j=\sum_{j=0}^{\infty}(j+1)a_{j+1}x^j,$$

where in the last equality we drop the term with $j=-1$ as it is identically zero. Applying this concept twice on the last term of (A.42c) and then relabeling $j=k$ in the result gives us equation (A.42d).

[7]Or 'orthogonal'. The proof of this property of the polynomial functions can be found in many analysis and linear algebra books.

[8]Except, of course, if you choose $\lambda=n(n+1)$.

n	$P_n(x)$
0	1
1	x
2	$\frac{1}{2}(3x-1)$
3	$\frac{1}{2}(5x^3-3x)$
4	$\frac{1}{8}(35x^4-30x^2+3)$
5	$\frac{1}{8}(63x^5-70x^3+15x)$

Table A.1: The first five (and zeroth, for good measure) Legendre polynomials, the solutions of equation A.40 for the given value of n and the appropriate choice of a_0 and a_1.

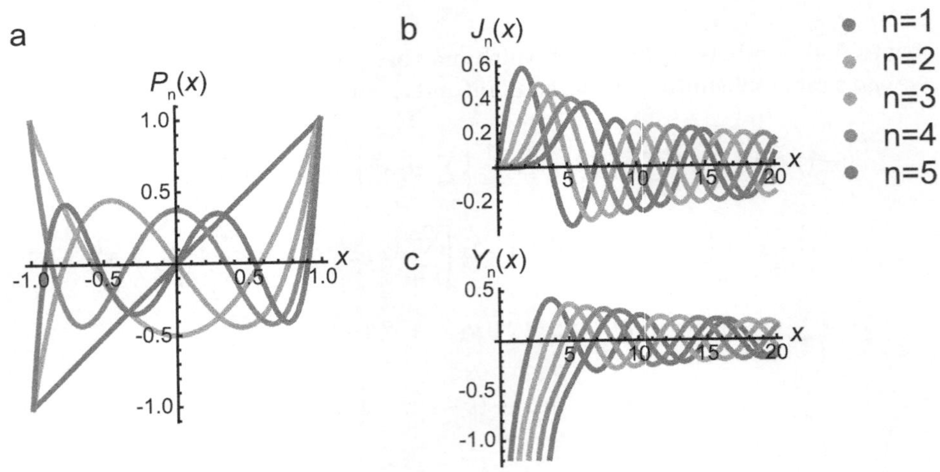

Figure A.4: Solutions to the Legendre and Bessel differential equations. (a) The first five Legendre polynomials (table A.1). Note that the polynomials with even n are all even, and those with odd n are all odd. (b-c) The first five Bessel functions of the first (b) and second (c) kind.

that lead to both infinite series and polynomial solutions. A well-known example is the Bessel differential equation:

$$x^2\frac{\mathrm{d}^2y}{\mathrm{d}x^2}+x\frac{\mathrm{d}y}{\mathrm{d}x}+(x^2-n^2)y(x)=0. \tag{A.44}$$

The solutions to this equation are known as the Bessel functions of the first and second kind (see problem A.3.3, where you'll prove that for these functions the series never terminates). These functions generalize the sine and cosine function and occur in the vibrations of two-dimensional surfaces. Other examples include the Hermite and Laguerre polynomials, which feature in quantum mechanics, and Airy functions, which you can encounter when studying optics.

A.3.6 PROBLEMS

A.3.1 (a) Suppose we have a solution of equation (A.24) where the roots $\lambda_\pm$ of the characteristic polynomial (equation A.26) are complex, so $\lambda_\pm = \alpha \pm i\beta$. Rewrite the general solution (A.27) in real functions with real coefficients C and D, and express C and D in terms of A and B. *Hint*: use Euler's formula $e^{ix} = \cos(x) + i\sin(x)$.

(b) Suppose we have a solution of equation (A.32) where the roots $n_\pm$ are complex, so $n_\pm = \alpha \pm i\beta$. To get a solution of equation (A.30) without complex numbers, we make the substitution $x = e^t$, so

$$x^{n_\pm} = x^{\alpha \pm i\beta} = e^{\alpha t} e^{\pm i\beta t}.$$

Use Euler's formula again to rewrite the complex exponential in terms of sines and cosines, and make the back-substitution to x to show that the general solution of equation (A.30) in this case is given by

$$y(x) = x^\alpha \left[A\cos\left(\beta \ln(x)\right) + \sin\left(\beta \ln(x)\right)\right]. \tag{A.45}$$

(c) Suppose we have a solution of equation (A.32) for which there is only a single root n. Show that the derivative of x^n with respect to n is in this case also a solution of equation (A.30), and that the general solution is given by

$$y(x) = x^n \left[A + B\ln(x)\right]. \tag{A.46}$$

A.3.2 Use the method of reduction of order to obtain a second solution of equation (A.24) for the case that the characteristic polynomial (equation A.26) has only a single root.

A.3.3 (a) Use the power series technique to find a solution to Bessel's differential equation (A.44). Why doesn't the series terminate in this case? Why do you only get one family of solutions? We'll call these solutions 'Bessel functions of the first kind' and label them as $J_n(x)$ (see figure A.4b).

(b) Use the method of reduction of order to find a second family of solutions to the Bessel differential equation, known as 'Bessel functions of the second kind' ($Y_n(x)$, see figure A.4c).

B

SOME EQUATIONS AND CONSTANTS

B.1 PHYSICAL CONSTANTS

Name	Symbol	Value
Speed of light	c	$3.00 \cdot 10^8$ m/s
Elementary charge	e	$1.60 \cdot 10^{-19}$ C
Electron mass	m_e	$9.11 \cdot 10^{-31}$ kg $= 0.511$ MeV/c^2
Proton mass	m_p	$1.67 \cdot 10^{-27}$ kg $= 938$ MeV/c^2
Gravitational constant	G	$6.67 \cdot 10^{-11}$ N·m^2/kg^2
Gravitational acceleration	g	9.81 m/s^2
Boltzmann's constant	k_B	$1.38 \cdot 10^{-23}$ J/K
Planck's constant	h	$6.63 \cdot 10^{-34}$ J·s
	$\hbar = h/2\pi$	$1.05 \cdot 10^{-34}$ J·s

Table B.1: Physical constants

B.2 MOMENTS OF INERTIA

Object	Moment of inertia
Thin stick (length L)	$\frac{1}{12}ML^2$
Ring of hollow cylinder (radius R)	MR^2
Disk or solid cylinder (radius R)	$\frac{1}{2}MR^2$
Hollow sphere (radius R)	$\frac{2}{3}MR^2$
Solid sphere (radius R)	$\frac{2}{5}MR^2$
Rectangle (size $a \times b$), perpendicular axis	$\frac{1}{12}M(a^2+b^2)$
Rectangle (size $a \times b$), axis parallel to side b	$\frac{1}{12}Ma^2$

Table B.2: Moments of inertia, all about axes of symmetry through the center of mass.

B.3 SOLAR SYSTEM OBJECTS

	Sun	Earth	Moon
Mass (kg)	$1.99 \cdot 10^{30}$	$5.97 \cdot 10^{24}$	$7.35 \cdot 10^{22}$
Mean radius (m)	$6.96 \cdot 10^{8}$	$6.37 \cdot 10^{6}$	$1.74 \cdot 10^{6}$
Orbital period (s)	$6 \cdot 10^{15}$	$3.16 \cdot 10^{7}$	$2.36 \cdot 10^{6}$
	(200 My)	(365.25 days)	(27.3 days)
Mean orbital radius (m)	$2.6 \cdot 10^{20}$	$1.50 \cdot 10^{11}$	$3.85 \cdot 10^{8}$
Mean density (kg/m^3)	$1.4 \cdot 10^{3}$	$5.5 \cdot 10^{3}$	$3.3 \cdot 10^{3}$

Table B.3: Characteristics of the Sun, Earth and Moon.

Name	Symbol	Equatorial radius	Mass	Mean orbit radius	Orbital period	Inclination	Orbital eccentricity	Rotation period	Confirmed moons	Axial tilt
Mercury	☿	0.382	0.06	0.39	0.24	3.38	0.206	58.64	0	0.04
Venus	♀	0.949	0.82	0.72	0.62	3.86	0.007	-243.02	0	177.36
Earth	♁	1	1	1	1	7.25	0.017	1	1	23.44
Moon	☾	0.272	0.0123	384399	27.32158	18.29-28.58	0.0549	27.32158	0	6.68
Mars	♂	0.532	0.107	1.52	1.88	5.65	0.093	1.03	2	25.19
Ceres		0.0742	0.00016	2.766	4.599	10.59	0.08	0.3781	0	4
Jupiter	♃	11.209	317.8	5.2	11.86	6.09	0.048	0.41	69	3.13
Io		0.285	0.015	421600	1.769	0.04	0.0041	1.769	0	0
Europa		0.246	0.008	670900	3.551	0.47	0.009	3.551	0	0
Ganymede		0.413	0.025	1070400	7.155	1.85	0.0013	7.155	0	0
Callisto		0.378	0.018	1882700	16.689	0.2	0.0074	16.689	0	0
Saturn	♄	9.449	95.2	9.54	29.46	5.51	0.054	0.43	62	26.73
Titan		0.404	0.023	1221870	15.945	0.33	0.0288	15.945	0	0
Uranus	⛢	4.007	14.6	19.22	84.01	6.48	0.047	-0.72	27	97.77
Oberon		0.119	0.00051	583519	13.46	0.1	0.0014	13.46	0	0
Neptune	♅	3.883	17.2	30.06	164.8	6.43	0.009	0.67	14	28.32
Triton		0.212	0.00358	354759	5.877	157	0.00002	5.877	0	0
Pluto	♇	0.186	0.0022	39.482	247.9	17.14	0.25	6.39	5	119.59
Charon		0.095	0.00025	17536	6.387	0.001	0.0022	6.387	0	unknown
Haumea		0.13	0.0007	43.335	285.4	28.19	0.19	0.167	2	unknown
Makemake		0.11	unknown	45.792	309.9	28.96	0.16	unknown	1	unknown
Eris		0.18	0.0028	67.668	557	44.19	0.44	unknown	1	unknown

Table B.4: Properties of a number of solar system objects. Equatorial radii and masses are compared to those of Earth (see table B.3). Orbital properties are around primary (the sun for (dwarf) planets, the planet for moons). Orbital radii and periods for planets again compared to Earth, for moons in kilograms and days. Rotation period for all objects in days. Inclination and axial tilt in degrees. Data from NASA planetary fact sheets [31].

B.4 EQUATIONS

B.4.1 VECTOR DERIVATIVES

Gradient:

$$\nabla f(\boldsymbol{r}) = \nabla f(x,y,z) = \begin{pmatrix} \partial_x f \\ \partial_y f \\ \partial_z f \end{pmatrix} = \left(\frac{\partial f}{\partial x}\hat{x} + \frac{\partial f}{\partial y}\hat{y} + \frac{\partial f}{\partial z}\hat{z} \right). \tag{B.1}$$

Divergence:

$$\nabla \cdot \boldsymbol{v} = \left(\partial_x, \partial_y, \partial_z\right) \cdot \begin{pmatrix} v_x \\ v_y \\ v_z \end{pmatrix} = \frac{\partial v_x}{\partial x} + \frac{\partial v_y}{\partial y} + \frac{\partial v_z}{\partial z}. \tag{B.2}$$

Curl:

$$\nabla \times \boldsymbol{A} = \left(\partial_x, \partial_y, \partial_z\right) \times \begin{pmatrix} A_x \\ A_y \\ A_z \end{pmatrix} = \begin{pmatrix} \partial_y A_z - \partial_z A_y \\ \partial_z A_x - \partial_x A_z \\ \partial_x A_y - \partial_y A_x \end{pmatrix}. \tag{B.3}$$

B.4.2 SPECIAL RELATIVITY

Lorentz transformations for the coordinates of a frame S' that moves with a speed u in the positive x-direction of frame S:

$$x' = \gamma(u)\left(x - \frac{u}{c}ct\right) \tag{B.4}$$

$$ct' = \gamma(u)\left(ct - \frac{u}{c}x\right) \tag{B.5}$$

$$\gamma(u) = \frac{1}{\sqrt{1-(u/c)^2}} \tag{B.6}$$

Velocity addition in a relativistic system:

$$v_x = \frac{u + v'_x}{1 + uv'_x/c^2} \text{ (longitudinal)}, \qquad v_y = \frac{1}{\gamma(u)} \frac{v'_y}{1 + uv'_x/c^2} \text{ (transversal)}. \tag{B.7}$$

C

IMAGE, DATA AND PROBLEM CREDITS

C.1 IMAGES

1. Close-up of the Prague astronomical clock by Fabrizio Verrecchia, from Unsplash, public domain.
2. Portrait of Isaac Newton by Godfrey Kneller, painted in 1689. Picture is a faithful reproduction of a two dimensional public domain work, retrieved from Wikimedia commons.
3. Drawing by Robert Hooke of the cell structure of cork from two different angles, as published in his 1665 book Micrographia. Picture is a faithful reproduction of a two dimensional public domain work, retrieved from Wikimedia commons.
4. Portrait of Galileo Galilei by Justus Sustermans, painted in 1636, currently located at the National Maritime Museum in London, UK. Picture is a faithful reproduction of a two dimensional public domain work, retrieved from Wikimedia commons.
5. Portrait of Charles de Coulomb. Picture is a faithful reproduction of a two dimensional public domain work, retrieved from Wikimedia commons.
6. Images by Michael Maggs, Wikimedia commons, falling ball and Wikimedia commons, bouncing ball, CC BY-SA 3.0.
7. Image by Amber Turner, Unsplash, public domain.
8. Image by GSenkow, Wikimedia commons, CC BY-SA 3.0.
9. Image cropped from fruit stand image by Redd Angelo, Unsplash, public domain.
10. Image by Malene Thyssen, Wikimedia commons, CC BY-SA 3.0.
11. Image of Allyson Felix at the 2012 Summer Olympics by Citizen59, Wikimedia commons, CC BY-SA 3.0.
12. Image of Robert Garrett throwing the discus at the 1896 Summer Olympics; picture by Albert Meyer, shared by the Bulgarian Archives State Agency (file BASA-3K-7-422-22) through Wikimedia commons, public domain.
13. Photograph of Emmy Noether, unknown date and photographer, obtained from Wikimedia commons, public domain.
14. Photograph of Konstantin Eduardovich Tsiolkovsky, unknown date and photographer, obtained from Wikimedia commons, public domain.
15. Images from NASA, public domain.
16. Image by Securiger, Wikimedia commons, CC BY-SA 3.0.
17. Image by Jayess, Wikimedia commons, public domain.

18. Image by Markmcgee, Wikimedia commons, public domain.
19. Image by Arnaud 25, Wikimedia commons, public domain.
20. Image from NASA's Aqua/MODIS satellite, retreived from Wikimedia commons, public domain.
21. Portrait of Johannes Kepler by an unknown artist, 1610. Picture is a faithful reproduction of a two dimensional public domain work, retrieved from Wikimedia commons.
22. Portrait of Leonhard Euler by Jakob Emanuel Handmann, 1753. Picture is a faithful reproduction of a two dimensional public domain work (now in the Kunstmuseum Basel), retrieved from Wikimedia commons.
23. Portrait of Christiaan Huygens by Caspar Netscher, painted in 1671, currently located at museum Boerhave in Leiden, The Netherlands. Picture is a faithful reproduction of a two dimensional public domain work, retrieved from Wikimedia commons.
24. Photograph of Christian Doppler, unknown date and photographer, obtained from Wikimedia commons, public domain.
25. Mach's 1888 photograph of the shockwaves created by a supersonic brass bullet. This photo was taken in Prague, using Schlieren Photography on a 5 mm-diameter negative. Obtained from Wikimedia commons, public domain.
26. Photograph of Ernst Mach from the Zeitschrift für Physikalische Chemie, Band 40, 1902, obtained from Wikimedia commons, public domain.
27. Official 1921 Nobel prize photograph of Albert Einstein, obtained from Nobelprize.org, public domain.
28. Photograph of Lorentz from the collection of Museum Boerhaave, Leiden, The Netherlands, obtained from Wikimedia commons, public domain.
29. Photograph of Minkowski from the book *Raum und Zeit* (Jahresberichte der Deutschen Mathematiker Vereinigung, Leipzig, 1909), obtained from Wikimedia commons, public domain.
30. Screen shots from buienradar.nl, taken on October 28, 2013.

C.2 DATA

31. Data from NASA planetary fact sheets at https://nssdc.gsfc.nasa.gov/planetary/planetfact.html.

C.3 PROBLEMS

32. Problem inspired by problem 4-s1 from 'Problems in introductory physics' by Crowell and Shotwell, available on Light and matter, CC BY-SA 3.0.

D

Summary and author biograpy

D.1 Summary

In *Mechanics and Relativity*, the reader is taken on a tour through time and space. Starting from the basic axioms formulated by Newton and Einstein, the theory of motion at both the everyday and the highly relativistic level is developed without the need of prior knowledge. The relevant mathematics is provided in an appendix. The text contains various worked examples and a large number of original problems to help the reader develop an intuition for the physics. Applications covered in the book span a wide range of physical phenomena, including rocket motion, spinning tennis rackets and high-energy particle collisions.

D.2 About the author

Dr. **T. (Timon) Idema** is an associate professor at the Department of Bionanoscience at Delft University of Technology (TU Delft) in The Netherlands. Before starting his research group in Delft in 2012, Idema obtained his PhD in theoretical biophysics at Leiden University (The Netherlands) and worked at the Institut Curie (Paris, France) and the University of Pennsylvania (Philadelphia, USA).

Idema's group studies collective dynamics in biologically motivated systems, ranging from proteins at the nano scale to tissues and even populations at the micro- and macro scale. A theorist himself, Idema frequently collaborates and co-publishes with experimental groups. He also teaches a number of courses at TU Delft, ranging from introductory physics to courses on soft matter and geometry that take students to the cutting edge of current research. For more details on his group's research and teaching activities, visit their website at idemalab.tudelft.nl.

INDEX

9 788197 559419

Printed at Repro India Ltd.